Microbial Food Safety

A Food Systems Approach

Microbial Food Safety

A Food Systems Approach

Charlene Wolf-Hall

Department of Veterinary and Microbiological Sciences,
North Dakota State University,
USA

and

William Nganje

Department of Agribusiness and Applied Economics,
North Dakota State University,
USA

CABI is a trading name of CAB International

CABI	CABI
Nosworthy Way	745 Atlantic Avenue
Wallingford	8th Floor
Oxfordshire OX10 8DE	Boston, MA 02111
UK	USA

Tel: +44 (0)1491 832111
Fax: +44 (0)1491 833508
E-mail: info@cabi.org
Website: www.cabi.org

Tel: +1 (617)682-9015
E-mail: cabi-nao@cabi.org

A catalogue record for this book is available from the British Library, London, UK.

Library of Congress Cataloging-in-Publication Data

Names: Wolf-Hall, Charlene, author. | Nganje, William Evange, 1966- author.
Title: Microbial food safety : a food systems approach / Charlene Wolf-Hall and William Nganje.
Description: Wallingford, Oxfordshire, UK ; Boston, MA : CABI, [2017] |
   Includes bibliographical references and index.
Identifiers: LCCN 2016042167 (print) | LCCN 2016043083 (ebook) | ISBN
   9781780644806 (hardback : alk. paper) | ISBN 9781780644813 (pbk. : alk.
   paper) | ISBN 9781780644820 (pdf) | ISBN 9781786391971 (ePub)
Subjects: | MESH: Food Contamination--prevention & control | Food
   Contamination--economics | Food Microbiology | Foodborne
   Diseases--economics | Systems Analysis
Classification: LCC RC143 (print) | LCC RC143 (ebook) | NLM WA 701 | DDC
   363.19/2--dc23
LC record available at https://lccn.loc.gov/2016042167

ISBN-13: 978 1 78064 480 6 (hbk)
          978 1 78064 481 3 (pbk)

Commissioning editor: Rachel Cutts and Rachael Russell
Associate editor: Alexandra Lainsbury
Production editor: James Bishop

Typeset by SPi, Pondicherry, India
Printed and bound in the UK by CPI Group (UK) Ltd, Croydon, CR0 4YY

Contents

This book is enhanced with supplementary resources.
To access the customizable lecture slides please visit:
ww.cabi.org/openresources/44813

1 Food

Key Questions
- What are some of the major food groups consumed globally?
- What are some examples of unusual or exotic foods?
- What are food supply networks?

In the field of food safety, common terminology used includes risk and risk management to prevent foodborne illnesses caused by microbial, chemical, or physical hazards. This chapter will focus on the complexity that is food and introduce concepts for how microorganisms can become problems or risks in foods. It will also introduce thinking about how humans manage the food supply to lower the risk of foodborne illness caused by microorganisms.

Review of Biological Kingdoms and Who is Eating Whom

To understand food systems, one should first thoroughly understand food. Everyone thinks they understand food, since we have been eating our entire lives. You are familiar with the foods that you purchase, and perhaps even grow in your own garden. You may even understand how food is produced and processed. Let's take some time to review what food is in the biological sense.

Food, as defined by the Oxford English Dictionary, is "Any nutritious substance that people or animals eat or drink in order to maintain life and growth; nourishment, provisions." Basically, food is a source of energy and structural maintenance in living organisms. Any living organism can be a potential food source to other living organisms, humans included.

When we think about living organisms and food, we start thinking about plants and animals and the food substances they provide. But, think further.

Do we just consume plants and animals? Do you like mushrooms? When you eat a fresh salad, is it just plants?

For the purpose of our discussion, let's consider the five biological kingdoms: Animalia (animals), Plantae (plants), Fungi (yeasts, molds and mushrooms), Protista (protozoa), and Eubacteria (bacteria of importance in food). Also, keep in mind that biological organisms live in complex biological communities or ecosystems where it is very common to find multiple kingdoms represented. Most organisms used as food sources are not just that organism, but also the ecosystem they carry with them, most of which is microscopic. This microscopic environment is known as the microbiome.

In that fresh salad, there may be several kingdoms represented in its unique ecosystem (Fig. 1.1). In addition to all of these life forms, there may also be viruses (tiny nucleic acid-based organisms that are dependent on host cells for biological activity), which may be contaminants or infecting the other organisms or tissues in the bowl (Forterre, 2010). Once a human consumes the salad, some of the intact and viable microscopic organisms could remain and colonize the intestinal tract, and use the human host for food. This host utilization may harm (pathogenic/parasitism), benefit (probiotic/mutualism), or make no significant difference (commensalism) to the host. Who is eating whom?

An old joke:
Q: What is worse than finding a worm in an apple?
A: Finding only half a worm!

Fig. 1.1. What is in that fresh salad? See Table 1.1 for more details.

Table 1.1. Biological kingdoms represented as tissues or intact in a fresh salad. See Fig. 1.1 for visual details.

Biological kingdom	Parts of the salad
Plants	Lettuce and any other fruits, vegetables, tofu, seeds, oils, and nuts
Animals	Meat, cheese, eggs, insects, mites (see Fig. 1.2), and worms
Fungi	Mushrooms, molds, and yeasts
Protozoa	Various harmless protozoa and possibly some parasites
Bacteria	Natural microflora of ingredients and possible pathogenic contaminants

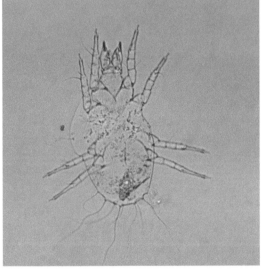

Figure 1.2. A photomicrograph of a female ham mite, *Tyrophagus puterscentiae*, an agricultural pest that infests cheeses and dried meats. (CDC Public Health Image Library/Margaret A. Parsons, available at: http://phil.cdc.gov/phil/details.asp?pid=3807.)

Typically, the more processed a human food is, the less living microscopic biological diversity is represented, but there may also be less nutrition. Only a few rare food items would be completely sterile and void of microscopic life. Foods considered commercially sterile and shelf stable can still harbor viable microbial spores. These are aspects of our food supply that are important to consider in complex food networks and in microbial food safety.

Let's explore some more about the main types of foods consumed by humans around the world to ensure we are thoroughly considering food systems. Kearney (2010) categorizes foods evaluated for global consumption trends into these categories:

- cereals;
- meat;
- eggs, milk, and other dairy products;
- fish;
- vegetables; and
- energy providers: vegetable oils, animal fats, and sugar.

The next few sections will relate to these terms for global consumption rates, but are categorized differently to stick to our biological contexts.

Animal-Based Foods

There is a wide array of animals that have supplied food for humans since the dawn of man. Many diseases of animals can be transmitted to humans through animal husbandry or food transmission. Zoonotic diseases are infectious diseases that are transmitted between vertebrate animals and humans (see Box 1.1).

Terrestrial mammals and birds

Animals, including mammals and birds that walk the land, have always been common sources of food for humans. Globally, the most common mammals eaten for meat by humans are goats, followed by sheep, pigs, and cattle; the most common birds are chickens (Kearney, 2010) (Fig. 1.3).

For a disturbing yet fascinating perspective on human culture and the history of animal use for food, readers are encouraged to read *Near a Thousand Tables* by Fernandez-Armesto (2002). This book includes a chapter describing cannibalism, and explains aspects of survival, culture, and aggression. This quote is used to start that chapter:

> Cannibalism is a problem. In many cases the practice is rooted in ritual and superstition rather than gastronomy, but not always. A French Dominican in the seventeenth century observed that the Caribs had most decided notions of the relative merits of their enemies. As one would expect, the French were delicious, by far the best. This is no surprise, even allowing for nationalism. The English came next, I'm glad to say. The Dutch were dull and stodgy and the Spaniards so stringy they were hardly a meal at all, even boiled. All this sounds sadly like gluttony.
>
> *Patrick Leigh Fermor*

Freshwater and seafood animals

The biological diversity within this category of animal foods is vast, and includes vertebrates and invertebrates. Global consumption trends include mainly white fish, oily fish, and seafood invertebrates (Kearney, 2010) (Fig. 1.4).

Box 1.1. An example of *Salmonella* as a cause of zoonotic disease

Salmonella are bacteria that can be found in many animal species from arthropods to large mammals. Not all cause disease in animals, and often animals, including humans, are carriers. There are many strains of *Salmonella*, and Chapter 7 will go into more detail about these. Of these many strains, the ones that have been linked to animal infection have often been directly correlated to specific animal species, and are considered host adapted. However, several of these host-adapted strains can also cause illness in humans. When these strains are transmitted from animal to human, or vice versa, that is considered a zoonosis.

Salmonella is most commonly transferred through the fecal–oral route, although eggs and milk also serve as sources of disease-causing *Salmonella*. The bacteria are able to colonize the internal portion of the egg as it develops in the ovary of an infected animal or the milk-secreting tissues of an infected mammal. When a human consumes an undercooked egg, non-pasteurized milk, or fecal-contaminated food, salmonellosis can be the result. According to Forshell and Wierup (2006) "Salmonellosis is the most common food-borne bacterial disease in the world."

Fig. 1.3. The most common terrestrial food mammals in order of global consumption rates (left to right, top to bottom). Creative Commons courtesy of photographers: InspireFate Photography, G =], United Soybean Board, and Pamela, respectively. Available at: www.flickr.com.)

Insects, annelids, and reptiles

Data are lacking for global consumption rates of these animals. However, in certain parts of the world, these can provide a critical source of protein.

"Approximately 1,900 insect species are eaten worldwide, mainly in developing countries" (van Huis, 2013). The potential for insects as mini-livestock to mitigate environmental impact and food sustainability has been reviewed by van Huis (2013). Likely candidates include yellow mealworm, the house cricket, and the migratory locust. The Food and Agriculture Organization of the United Nations encourages insect consumption as part of a strategy to ensure an adequate world food supply, and state:

> Insects as food and feed emerge as an especially relevant issue in the twenty-first century due to the rising cost of animal protein, food and feed insecurity, environmental pressures, population growth and increasing demand for protein among the middle classes. Thus, alternative solutions to conventional

livestock and feed sources urgently need to be found. The consumption of insects, or entomophagy, therefore contributes positively to the environment and to health and livelihoods (van Huis *et al.*, 2013).

Readers can view an interesting take on this through the National Geographic website (National Geographic, 2016).

Reptile consumption is more accepted globally than insect consumption. Turtle soup is a popular delicacy, and has resulted in threats to green turtle species (Klemens and Thorbjarnarson, 1995). Reese (1917) described a variety of reptiles that could be consumed by Americans.

Readers may have had an early literary influence regarding consumption of annelids through the children's book by Thomas Rockwell entitled *How to Eat Fried Worms*. Sabine (1983) wrote a chapter on "Earthworms as a source of food and drugs," and concluded that earthworms could be considered as high-protein food for humans or animals, but cautioned regarding food safety

Fig. 1.4. Examples of seafood invertebrates, otherwise known as shellfish. (Dahlström, 2013; available at: https://www.flickr.com/photos/dahlstroms/with/8637094579)

hazards such as accumulated heavy metals (e.g. lead) and agrochemicals (e.g. pesticides).

Dairy products and eggs

These foods could be considered as edible excretions from animals. Their biological function is to provide food for offspring, and are therefore highly nutritious. Dairy products and eggs provide a renewable protein source, and are important dietary staples around the world.

Dairy products come from milk, which can come from any mammalian source. Most of humankind starts off life consuming human milk. The most common agricultural source of milk globally is cow's milk and its various processed forms, like yogurt and cheese. Other sources of milk for human consumption include buffalo, goats, sheep, camels, horses, donkeys, reindeer, and yaks, with the first four producing 11%, 2%, 1.4%, and 0.2%, respectively, of all milk worldwide in 2011 (Gerosa and Skoet, 2012).

Eggs can come from any animal including birds, reptiles, amphibians, fish, and insects. Mammals also produce eggs. The platypus would be an example of a mammal that excretes or lays eggs. The chicken egg is the most commonly consumed type of egg worldwide. Other commonly consumed bird's eggs include duck, goose, quail, ostrich, and gull eggs. Fish eggs, or roe, are also commonly consumed worldwide. The very expensive roe of the sturgeon that has been salt cured is called caviar and is considered a delicacy around the world.

Plant-Based Foods

A major source of food for humans is the edible plants of the world. Plants have evolved as food sources for many creatures, and some of the evolutionary paths have resulted in poisonous plants. Fortunately, many plants have evolved to be edible as a means of carrying on their progeny. Humans have directed edible plant evolution since the dawn of agriculture.

Fruits and vegetables

Plant foods often are categorized in either botanical or culinary terms. In botanical terms, a fruit is a seed-bearing structure that develops from the ovary of a flowering plant, and vegetables are all other plant parts, including roots, leaves, and stems. In culinary terms, fruits are often categorized as sweet, whereas vegetables are considered savory. The tomato is a common example of a plant food that can be described as either a fruit or a vegetable. See Box 1.2 for information on how fruits and vegetables can become contaminated with a pathogen like *Salmonella*.

Nuts, grains, legumes, and oilseeds

The seeds of plants serve as food staples around the world. They contain high concentrations of energy, providing chemicals intended to feed the germination process for new plant growth. This characteristic also makes them a highly nutritious food source for humans and animals.

Microbial-Based Foods

There are many food products based on microorganisms, or for which microorganisms play a major role in production.

Mushrooms and other fungi

Mushrooms are commonly consumed globally, and are the fruiting bodies of certain types of fungi. Just as with plants, many have evolved to be poisonous, whereas others are edible and have been cultivated by humans. Other lesser known fungal-based foods have gained some popular status in pockets around the world. Australians may be familiar with the popular spread product known as Vegemite, a food product derived from yeast cells. Those in Europe may be familiar with the meat substitute product called Quorn, derived from a mold, also known as mycoprotein.

Single-cell protein

Single-cell protein (SCP) is a term used for protein derived from unicellular microorganisms (Snyder, 1970). Other names for the same type of material are novel protein, unconventional protein, minifoods, and petroprotein. This food material consists of the dead cells of fungi, algae, or bacteria. It is typically processed in a manner to reduce toxicity due to high concentrations of nucleic acids. This material has potential as human food or animal feed.

Advantages of SCP include fast production times from cheap substrates. Some algae in combination with nitrogen-fixing *Azotobacter* can produce protein from simple substrates such as water, sunlight, and air. Disadvantages include the necessity for pure cultures, transportation costs, risks as allergens, toxicity, and low digestibility of cell wall components.

> The greater speed and efficiency of microbial protein production compared to plant and animal sources may be illustrated as follows: a 1,000-lb steer produces about 1 lb. of new protein per day; soybeans (prorated over a growing season) produce about 80 lb., and yeasts produce about 50 tons (Jay *et al.*, 2005).

There are examples of SCP products on the market. These include mycoprotein-based products like Quorn and Vegemite. Have you ever had a *Spirulina* smoothie? Spirulina is a cyanobacterium.

Fermented foods

Food fermentation is a mechanism by which microorganisms are utilized to preserve or enhance food. There are many fermented foods in the world, and pretty much any form of food can be fermented. Entire books are available on the subject (Hutkins, 2006). Typically, the main microbial metabolic products produced for these purposes are lactic acid, alcohol, or acetic acid, although very complex reactions can happen as microorganisms pre-digest our food for us. Examples of foods that many do not realize are fermented include the flavorful chocolate, vanilla, coffee, tea, soy sauce, butter, and many more.

Multi-ingredient processed foods

Much of the food we eat is made up of multiple ingredients. This increases the complexity of our biological context for food by mixing together many ecosystems to form new ones. Just look at the food label on any item in your cupboard. The multiple components included may have come from many different places.

Food Supply Networks

Now that we have a biological perspective about food and the microscopic ecosystems that they include, let's consider how food gets to our tables.

Food has to move as not all humans have access to adequate food supplies. Current global circumstances have food distributed inequitably, with the result contributing to obesity and related chronic health issues in developed countries and malnutrition and starvation in others. Even in developed countries, there are places described as food deserts. Food deserts are communities with limited access to fresh fruit, vegetables, and other healthy foods. These could be impoverished areas, at times within major cities. These areas lack mainstream grocery stores, farmers' markets, and healthy food providers. In the United States, the United States Department of Agriculture defines a food desert as a "low-access community" where at least 500 people and/or at least 33% of the population reside more than 1 mile from a supermarket or large grocery store (for rural census tracts, the distance is more than 10 miles) (USDA ERS, 2015). Food deserts have become a major concern because these communities only have access to more processed foods, sugar, and fatty foods, which contribute to obesity and obesity-related diseases.

Inefficient food distribution contributes to global food insecurity. The Food and Agricultural Organization reported that almost 870 million people were chronically undernourished in 2012 (FAO, 2016). With the current rate of human population growth, it is projected that by 2050 the world's population will reach 9.1 billion, 34% higher than today. Food production must increase by 70% to accommodate the increased population. It is even more critical that food supply networks are developed to efficiently distribute food.

"Food supply chains stretch from agricultural producers to consumers and usually involve a manufacturing stage, as well as foodservice or retail activities" (Akkerman *et al.*, 2010). Food supply networks account for more complexity than a simple linear model for food distribution. Figure 1.5 provides a representation of some of the components that impact food supply networks.

Supply network complexity and spread of food hazards

The foods presented in the previous sections could be classified into four food supply network categories as depicted in Fig. 1.6: tightly coupled, linear supply system; tightly coupled, complex supply system; loosely coupled, linear supply system; and loosely coupled, complex supply system.

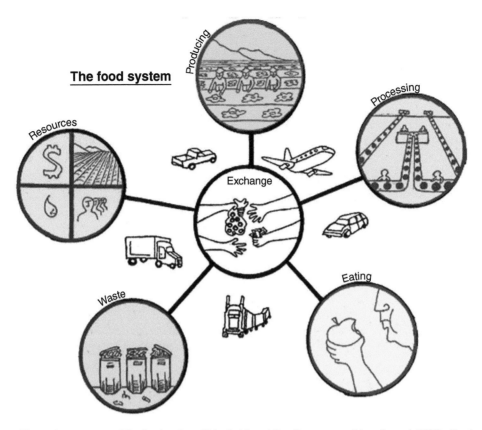

The food system

Producing

Processing

Resources

Exchange

Eating

Waste

Fig. 1.5. The various aspects of the food system. (Adapted from https://commons.wikimedia.org/wiki/File:Foodsystem.jpg; Creative Commons, author Hunt041.)

		Unpopulated region
Tight coupling	Fresh bagged leafy greens (easy trace)	Branded processed food products (mixed trace)
Loose coupling	Farmer's market, Ontario: raw milk and cheese event (medium to easy trace)	Tomatoes/peppers event (difficult trace)
	Linear network	Complex network

Fig. 1.6. Perrow's typology applied to food supply networks. (Adapted from Perrow, 1999.)

Perrow's typology (Perrow, 1999) would classify our salad bowl (Fig. 1.1) as an example of a loosely coupled, complex supply system (Fig. 1.6). The complexity of the salad bowl arises because complex interactions between foods from different sources multiply the opportunities for hard-to-understand food safety hazards to occur. Complex systems have unplanned, unexpected, invisible, unfamiliar, ambiguous, or incomprehensible sequences, often made more obscure by poorly understood transformation processes. Less complex systems are likely to be more traceable because interactions will be less complex, transformations will be fewer, and the number and variety of components or actors involved will be less. Linear systems afford a greater opportunity to get a clear picture of the whole system, making them easier to trace back. Perrow was also concerned with the degree of coupling between parts of systems. Loosely coupled systems allow for processing delays, do not have fixed sequences or relationships, retain slack resources, and exploit fortuitous substitution possibilities. In tightly coupled systems, performance standards are enforced and unambiguous. Delays are minimized, sequences are invariant, methods are constrained, any buffers or redundancies are designed, and substitutions or commingling are limited. Loosely coupled systems are less likely to be traceable because information is more likely to be ambiguous and less likely to be visible to and understood by the various actors in the system. Tightly coupled systems thus facilitate trace back by ensuring that information is captured, maintained, and readily available for rapid recall. Microbial food hazards are limited with tightly coupled, linear networks because the source of contamination or potential food hazard can be traced in a timely, targeted, and cost-effective manner.

Microbial food hazards, traceability, and recalls in food supply networks

There is little question that rapid response and targeted trace-back systems can minimize the economic damage inflicted by microbial food safety events by speeding up and narrowing product recalls. Faster recalls avoid additional cases of illness or death, and targeted recalls avoid false alarms on products that are safe. Despite this awareness and the proliferation of traceability standards and systems, investigators in food supply networks often find that when faced with an unexpected failure, participants "scramble" to produce the required information, leading to information losses or errors (Charlier and Valceschini, 2008) that turn into delays. One reason for these problems is that traceability systems are not uniformly implemented globally. However, PR Newswire (2014) reported that world revenue for food traceability technologies would reach US$10.59bn in 2014. Greater urgency from retailers to mandate suppliers to attach traceability information to food products stimulates investment in traceability. Several developed nations are taking the lead to mandate traceability requirements and technologies such as electronic barcode and radio-frequency identification systems. It is anticipated that increased investments in traceability will lead to reductions in foodborne illness outbreaks and recalls. Box 1.3 provides an example of the challenges to traceability.

Effective mitigation of food risks in food supply chains

Food supply networks are increasingly exposed to food safety and food defense risks partly due to the large volume of shipments from domestic and import sources (USDA AMS, 2012; FDA, 2014). Food safety can be defined as food system *reliability* – reducing exposure to natural hazards, errors, and failures. It is the unintentional contamination of food that may have dangerous and lingering consequences (FDA, 2014). Food defense, on the other hand, is system *resiliency* – reducing the impact of intentional system attacks either from disgruntled employees or terrorists. Either source of risk could be magnified as a result of error-based disruption (type I and II errors) or failures in prevention and control measures. A false positive or type I disruption occurs when an inspection system incorrectly identifies a threat, or a diagnostic system incorrectly identifies a cause, so a safe product is excluded from the supply chain. An example of a type I disruption would be a producer-initiated recall of finished products mistakenly believed to have been produced from contaminated raw materials or mistakenly believed to have been tampered with by terrorists.

A false-negative or type II disruption occurs when a defective product is distributed to the consumer and causes harm that is extensive enough to create market failure, as a result of a failure to detect the problem or diagnose the cause. Type II disruptions are associated with illness or death. Examples of type II disruptions are failures to detect accidental contamination from foodborne pathogens and adulteration. An example of type II disruption that will be used in several cases in this book is the 2008 jalapeño pepper outbreak in the United States, for produce imported from Mexico.

Economic consequences of food hazards in food supply chains

A single outbreak or food recall that causes illness and death could cost society hundreds of millions of dollars (Richards and Nganje, 2014). The length of time it takes to trace the source of the outbreak is positively correlated with the costs. The economic consequences are larger and more problematic for imported food with loose coupling and a complex network. For example, the 2006 California spinach *Escherichia coli* outbreak source was identified on 20 September, 47 days after the first reported case on 5 August, and the estimated cost was about US$127 million, while the 2008 *Salmonella enterica* outbreak of fresh jalapeño and serrano peppers from Mexico took 81 days and the reported costs were much larger.

Summary

Food is material that serves biological functions for living organisms. There are many types of foods with unique microbiomes in the global food supply chain. Vectors, fomites, movement, and mixing of foods through the food supply networks create many opportunities for transfer of microorganisms through food systems. Being able to trace the food supply chains back to sources can be a powerful tool to limit the number and severity of foodborne illnesses as well as reducing the economic impact.

Further Reading

Jay, J.M., Loessner, M.J., and Golden, D.A. (2005) *Modern Food Microbiology*, 7th edn. Springer, New York, USA.

Marchesi, J. (2014) *The Human Microbiota and Microbiome: Advances in Molecular and Cellular Microbiology*. CABI, Boston, MA, USA.

Pullman, M. and Wu, Z. (2012) *Food Supply Chain Management: Economic, Social and Environmental Perspectives*. Taylor & Francis, New York, USA.

References

Akkerman, R., Farahani, P., and Grunow, M. (2010) Quality, safety and sustainability in food distribution: a review of quantitative operations management approaches and challenges. *OR Spectrum* 32, 863–904.

Behravesh, C.B., Mody, R.K., Jungk, J., Gaul, L., Redd, J.T., Chen, S., Cosgrove, S., Hedican, E., Sweat, D., Chavez-Hauser, L., Snow, S.L., Hanson, H., Nguyen, T.A., Sodha, S.V., Boore, A.L., Russo, E., Mikoleit, M., Theobald, L., Gerner-Smidt, P., Hoekstra, R.M., Angulo, F.J., Swerdlow, D.L., Tauxe, R.V., Griffin, P.M., and Williams, I.T. (2011) 2008 outbreak of *Salmonella* Saintpaul infection associated with raw produce. *New England Journal of Medicine* 364, 918–927.

Berger, C.N., Sodah, S.V., Shaw, R.K., Griffin, P.M., Pink, D., Hand, P., and Frankel, G. (2010) Fresh fruit and vegetables as vehicles for the transmission of human pathogens. *Environmental Microbiology* 12, 2385–2397.

Charlier, C. and Valceschini, E. (2008) Coordination for traceability in the food chain. A critical appraisal of European regulation. *European Journal of Law and Economics* 25, 1–15.

FAO (2016) Globally almost 870 million chronically undernourished – new hunger report. Food and Agriculture Organization of the United Nations. Available at: http://www.fao.org/news/story/en/item/161819/icode/ (accessed 1 March 2016).

FDA (2008) *Salmonella* Saintpaul outbreak. United States Food and Drug Administration. Available at: http://www.fda.gov/NewsEvents/PublicHealthFocus/ucm179116.htm (accessed 11 September 2014).

FDA (2014) CARVER + Shock primer. United States Food and Drug Administration. Available at: http://www.fda.gov/Food/FoodDefense/FoodDefensePrograms/ucm376791.htm (accessed 30 December 2015).

Fernandez-Armesto, F. (2002) The meaning of eating: food as rite and magic. In: *Near a Thousand Tables: A History of Food*. The Free Press, New York, USA, pp. 21–54.

Forshell, L.P. and Wierup, M. (2006) *Salmonella* contamination: a significant challenge to the global marketing of animal food products. *Scientific and Technical Review of the Office International des Epizooties* 25, 541–554.

Forterre, P. (2010) Defining life: the virus viewpoint. *Origins of Life and Evolution of Biospheres* 40, 151–160.

Gerosa, S. and Skoet, J. (2012) *Milk Availability – Trends in Production and Demand and Medium-term Outlook*. ESA Working Paper No. 12-01. Food and Agriculture Organization of the United Nations, Rome, Italy.

Gourabathini, P., Brandl, M.T., Redding, K.S., Gunderson, J.H., and Berk, S.G. (2008) Interactions between foodborne pathogens and protozoa isolated from lettuce and spinach. *Applied and Environmental Microbiology* 74, 2518–2525.

Hutkins, R.W. (2006) *Microbiology and Technology of Fermented Foods*. Wiley-Blackwell, Ames, Iowa, USA.

Jay, J.M., Loessner, M.J., and Golden, D.A. (2005) Miscellaneous food products. In: *Modern Food Microbiology*, 7th edn, Springer, New York, USA, pp. 197–213.

Kearney, J. (2010) Food consumption trends and drivers. *Philosophical Transactions of the Royal Society B* 365, 2793–2807.

Klemens, M.W. and Thorbjarnarson, J.B. (1995) Reptiles as a food source. *Biodiversity and Conservation* 4, 281–298.

National Geographic (2016) Bugged out. Available at: http://channel.nationalgeographic.com/channel/doomsday-preppers/videos/bugged-out/ (accessed 1 March 2016).

Perrow, C. (1999) *Normal Accidents: Living with High Risk Technologies*. Princeton University Press, Princeton, NJ, USA.

PR Newswire (2014) Food traceability technologies market 2014–2024 – EAN/UPC Barcodes, 2D QR Codes, RFID & RTLS. Available at: http://www.marketwatch.com/story/smartech-publishing-report-says-google-glass-heralds-a-new-era-for-the-augmented-reality-business-2013-03-13 (accessed 22 May 2014).

Reese, A.M. (1917) Reptiles as food. *Scientific Monthly* 5, 545–550.

Richards, T. and Nganje, W. (2014) Welfare effects of food recall. *Canadian Journal of Agricultural Economics* 62, 107–124.

Sabine, J.R. (1983) Earthworms as a source of food and drugs. In: *Earthworm Ecology*. Springer, The Netherlands, pp. 285–296.

Snyder, H.E. (1970) Microbial sources of protein. *Advances in Food Research* 18, 85–140.

USDA AMS (2012) Microbiological Data Program. United States Department of Agriculture, Agricultural Marketing Service. Available at: https://www.ams.usda.gov/datasets/mdp (accessed 30 December 2015).

USDA ERS (2015) Food Access Reasearch Atlas. United States Department of Agriculture, Economic Research Service. Available at: http://www.ers.usda.gov/data-products/food-access-research-atlas.aspx (accessed 30 December 2015).

van Huis, A. (2013) Potential of insects as food and feed in assuring food security. *Annual Review of Entomology* 58, 563–583.

van Huis, A., van Itterbeeck, J., Klunder, H., Merens, E., Halloran, A., Muir, G., and Vantomme, P. (2013) *Edible Insects: Future Prospects for Food and Feed Security*. FAO Forestry Paper 171. Food and Agriculture Organization of the United Nations, Rome, Italy.

2 Ecological Concepts of Foods and Definitions of Pre- and Post-Harvest

> **Key Questions**
> - What is ecology, and how are these concepts applied to foods?
> - How are microorganisms affected by the environmental factors in foods?
> - What is hurdle technology?
> - What do pre- and post-harvest mean?

Introduction to Ecological Concepts of Food Microbiomes

In Chapter 1, we discussed food in terms of biological systems and food supply chains. In this chapter, we will discuss the environmental impacts on the internal microbiological systems of foods.

The Oxford English dictionary defines ecology as "The branch of biology that deals with the relations of organisms to one another and to their physical surroundings." There are many subcategories of specializations within the field of ecology, including the study of biomes. A biome is a naturally occurring community of biological organisms that can be classified into a specific group. Microbiomes are more specific to the microorganisms in a particular environment. Some may be familiar with the study of the human microbiome. Ecosystems of foods are microbiomes specific to individual foods.

Remember the fresh salad depicted in Chapter 1 (Fig. 1.1)? That diversity of biological kingdoms also represents a diversity of microbiomes. The microbiome of leaves is referred to as the phyllosphere, and leaves represent the largest biological surface on Earth (Remus-Emsermann and Vorholt, 2014). Plant materials such as lettuce leaves, tomato slices, and peppers, each carry their own microbial load of microorganisms. The same goes for the other ingredients in the salad. Nothing in that salad bowl is sterile or without some form of viable microorganism present. The specific microorganisms for

each are a result of the environments that those ingredients have been exposed to, and those ingredients will have been exposed to many environments from production, storage, processing, retail, and kitchen to table (Box 2.1). Our food is not sterile, and most microorganisms are harmless.

Biological Activity of Microorganisms in Foods

Experts in microbial food safety need to understand the basic biological functions of the microbial hazards of concern to implement effective control measures. Not all microorganisms are the same, and some are of concern after growth in a food, by just being present in a food, or from their chemical excretions that remain after they are dead.

Growth

Those who have studied basic microbiology will recall the concepts of a bacterial growth curve. The phases of a bacterial growth curve for bacteria inoculated into a new environment are depicted in Fig. 2.1. The lag phase is where multiplication has not started yet, but enzyme systems are working to adapt to the new environment and nutrients. The log or exponential phase is where the bacteria multiply at the most rapid rate. The stationary phase is where there is equilibrium between multiplication

Salmonella are not normal microflora on plant tissues, and their presence in raw produce is indicative of fecal contamination from animal sources. *Salmonella* can be transmitted through the food production chain to consumers. A large outbreak of salmonellosis occurred in 2008 (MMWR, 2008) and jalapeño peppers were indicated as the likely source. We may never know exactly how the *Salmonella* got into the peppers. Once in the peppers, there wasn't much that could be done to make the peppers safe to consume in raw form.

Liao *et al.* (2010) were able to demonstrate via inoculation studies that *Salmonella* Saintpaul could colonize and survive in jalapeño peppers. They concluded that the highest concentration of these bacteria was found in the stem/calyx, with less than 10% in other fleshy tissues. Those attached to the stem/calyx were more resistant to chemical sanitizing solutions. The *Salmonella* remained alive during refrigeration for 8 weeks, but did not increase in numbers.

At 68°C, the *Salmonella* were able to increase by 3 log units after 48 hours.

Pao *et al.* (2012) determined that *Salmonella* populations could rebound after washing treatments if kept under humid storage conditions at temperatures of 21°C or higher. Storage temperatures at 10°C or lower were adequate for preventing *Salmonella* from rebounding. It was recommended that uninterrupted cold storage would help lower the risk. This is consistent with the study by Castro-Rosas *et al.* (2011) who also recommended continuous temperature control to prevent growth. They also emphasized that this may not be adequate to eliminate the risk from virulent *Salmonella* strains that can cause infection at low infective doses.

Ecological factors considered in these types of studies include the type of strain, survival, growth, time, temperature, and stressors such as naturally present competitive microflora and physical or chemical sanitation treatments.

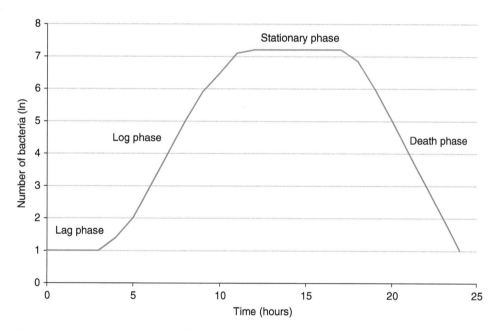

Fig. 2.1. The components of a generic bacterial growth curve.

and death. The death phase is where the resources of the environment have run out or wastes have built up, and the rate of death is higher than that of multiplication.

It is important to remember that the kinetics or basic patterns of a bacterial growth curve do not necessarily apply to other types of microorganisms such as viruses, fungi, or protozoa, which do not replicate via patterns of binary fission. Parasitic microorganisms, like viruses and some pathogenic protozoa, require a living host for replication and will not multiply in food (unless the food is a living host). Fungi may propagate from conidia or mycelia, so growth patterns can vary tremendously.

An understanding of growth patterns of specific microorganisms can be applied in mathematical modeling systems to account for changes in ecological systems of the microbial environment. These models can help predict if growth will occur in particular food systems, and can be valuable tools for food safety risk management (Box 2.2). According to Pin *et al.* (2013):

> During the last 30 years, predictive microbiology has developed into a discipline in its own right. This has been facilitated by dramatic developments in information technology (IT) and numerical and computational methods, which have revolutionised life sciences primarily via the new tools developed within bioinformatics and systems biology.

Freely available predictive modeling programs are online. Here are some example websites with more information:

- The Computational Microbiology Research Group – http://www.ifr.ac.uk/safety/comicro/default.html;
- The United States Department of Agriculture Agricultural Research Service Pathogen Modelling Program Online – http://ars.usda.gov/Services/docs.htm?docid=6786;
- Joint FAO/WHO Export Meetings on Microbiological Risk Assessment Risk Management Tool for the Control of Campylobacter and *Salmonella* in Chicken Meat – http://www.fstools.org/poultryRMTool/; and

- National Food Institute, Denmark, Food Spoilage and Safety Predictor (FSSP) – http://fssp.food.dtu.dk/default.aspx.

Survival

Although the ability of microorganisms to grow in a food can dramatically impact risk for food safety considerations, growth is not always a prerequisite to risk of foodborne disease. Some microorganisms can cause harm to a consumer at very low numbers, and their mere presence in a food can be an unacceptable risk. This is true for highly pathogenic bacteria, viruses, and parasites.

The term infective dose is used to describe the quantity, in terms of numbers, of a pathogen that is necessary to cause infection in a susceptible host. The United States Food and Drug Administration publishes the "Bad Bug Book" (FDA, 2012), which contains infective dose information for many foodborne pathogens. Infective doses as low as ten or fewer cells, cysts, worms, or virus particles are listed for particular strains of *Salmonella*, *Shigella*, *Coxiella burnetii*, *Francisella tularensis*, *Giardia lamblia*, *Entamoeba histolytica*, *Trichinella* spp., *Taenia* spp., *Anasakis simplex*, *Eustrongylides* spp., and noroviruses. For these pathogens, survival in a food is all that may be needed for successful transmission.

For other microorganisms, surviving food processing practices can lead to better opportunities for growth later. For example, bacteria that form heat-resistant spores, such as species of *Clostridium* and *Bacillus*, may survive heat processing and be able to germinate and begin multiplying as a more favourable temperature is reached. Chapter 3 will

---

Box 2.2. Modeling – not so easy for *Salmonella* in produce

The ecological characteristics of produce can range widely, and *Salmonella* strains do not all behave the same way. Considering these facts, there are still common patterns that can be modeled for risk assessment.

Pan and Schaffner (2010) noted that models for *Salmonella* in chickens were not that different from a model generated for cut red tomatoes as a function of time, even with very different pH values. In comparison to lab-generated data, two other modeling systems compared resulted in under-predicting

growth at lower temperatures or over-predicting growth at higher temperatures. This study points out some important limitations of models.

There is a need to study the impact of different control measures for *Salmonella* survival and growth in produce and other foods. Bermudez-Aguirre and Corradini (2012) point this out and emphasize the need for models to incorporate unique patterns that result from emerging intervention/control technologies. Modeling for *Salmonella* growth remains limited for predictability.

explain biological advantages that different food-borne microorganisms of concern can have for survival in food supply networks.

Secondary metabolites and microbial toxins

Microbes that cause infections are a major food safety concern, but so are bioactive chemicals produced by microorganisms that cause intoxication. Food safety experts clearly understand the different risk factors for infection versus intoxication. A foodborne infection requires ingestion of living microorganisms, typically at high enough numbers to establish an infection, numbers otherwise known as an infective dose. An intoxication is the result of ingesting biologically active chemicals that cause harm. We will revisit these concepts again later in Chapter 5.

Some microbial toxins can remain in foods long after the producing microbe is dead, and some are heat stable and remain biologically active after food processing. Most of these toxins are secondary metabolites. Primary metabolites are biologically produced molecules common in all biological kingdoms, and are necessary for energy metabolism, growth, development, and reproduction. Secondary metabolites are not produced by all organisms, are very diverse, tend to have biological activity, and are not critical to growth, development and reproduction. Although not needed for basic biological functions, secondary metabolites often provide advantages for surviving in what can be hostile environments. For example, venomous animals, poisonous plants and toadstools, and toxic microorganisms all possess advantages for obtaining food and preventing predation. Secondary metabolism in microorganisms can be triggered by depletion of a nutrient, the presence of an inducer, or as a result of a growth rate decrease as occurs during the stationary phase (Demain, 1998).

A group of important secondary metabolites produced by microorganisms is antibiotics. In common with toxins, antibiotics are biologically active but are purposely selected for activity against pathogenic bacteria with limited toxicity to humans or animals at controlled dosages. The story of the discovery and use of antibiotics is interesting and has been reviewed by others (Berdy, 2005).

Readers are encouraged to further explore the concepts of food toxicology to better understand the risks of biologically active compounds in foods, whether of microbial origin or from other sources.

Impact of Environmental Factors on Microorganisms in Foods

Understanding the environmental factors and how they impact microorganisms is important for dealing with microbial food safety.

Defining environmental factors in foods

An easy-to-remember acronym used by food safety trainers (NRAEF, 2014) for training food service workers is FAT TOM, for which:

F = Food
A = Acidity
T = Time
T = Temperature
O = Oxygen
M = Moisture

The above acronym is effective for remembering key environmental factors that affect microbial growth in food. However, food safety experts need to understand additional factors and the interplay of these factors on the biological activity of microbial foodborne hazards. Chapter 3 will more broadly define and describe the intrinsic and extrinsic environmental factors that impact microorganisms in foods. These factors can be modeled for predicting risk and can be manipulated to control the risk of foodborne disease for microbial food safety.

Hurdle Technology

Controlling the biological activity of microorganisms is critical for an adequate, high-quality, and safe food supply. An understanding of how microorganisms will react in food systems, especially where environmental conditions change as food moves through food supply networks, is critical for effective control of risk for foodborne illness.

The hurdle effect (or concept) has been described:

> to include combination preservation techniques that span the whole range of possible effects, including some that are sequential, but many more that act in unison, and include the operation of factors that are intrinsic, extrinsic, process-based, and implicit to a particular food.... From an understanding of the hurdle effect, hurdle technology has been derived (Leistner and Gould, 2002).

Using multiple approaches to control microbial growth or survival can be very effective. Environmental factors play an important role in hurdle technology. Other controls like cooking or irradiation can help preserve foods by killing microorganisms, but the tradeoffs are effects on quality. Hurdle technology allows for better quality food products with a low risk of foodborne illness.

Let's consider how shifts in environmental conditions can affect microorganisms in foods. Using the example of a bacterial growth curve, how will a shift in conditions affect the lag phase? This is the most critical phase to control. "A phenomenon inherent to microbial kinetics is lag, which is typically observed as a delayed response of the microbial population to a (sudden) change in the environment" (Swinnen *et al.*, 2004).

What will a bacterial growth curve look like in a food system incorporating multiple hurdles to growth? Let's consider a hypothetical (imagined) effect on bacteria introduced into a food system. Figure 2.2 depicts what a growth curve might look like in a food that is held at a temperature of 30°C (86°F). The hurdle to overcome for growth in this case is time.

Figure 2.3 shows growth and reduction curves for *Escherichia coli* O157:H7, generated by a modeling program that adjusts for the change in environmental factors. The first graph depicts *E. coli* under rather optimal growth conditions. The hurdle to growth is time in this example, and the lag is only a few hours. The second graph is *E. coli* under the same conditions, except that the temperature has been lowered, as in refrigeration. Now temperature is an additional hurdle to time, and the lag is extended to hundreds of hours. The third graph adds the third hurdle of low pH, and the lag is roughly doubled. The final graph represents *E. coli* in a food system with multiple hurdles, a type of sausage with preservatives and fermented by good bacteria that compete for nutrients. The hurdles become overwhelming in this system, no lag occurs, and the bacteria decrease or die off over time. This demonstrates the power of control that manipulating environmental factors can have on microorganisms in food. Individual factors do not need to be put at the extremes if combined with other factors for multiple hurdles.

Pre- and Post-Harvest

Food systems include the food production chain, or in other words how food gets from the producers' farms to the consumers' tables. In the food production chain, the terms pre- and post-harvest refer to stages prior to and after harvest of food, respectively. For example, corn kernels in a field are pre-harvest, while

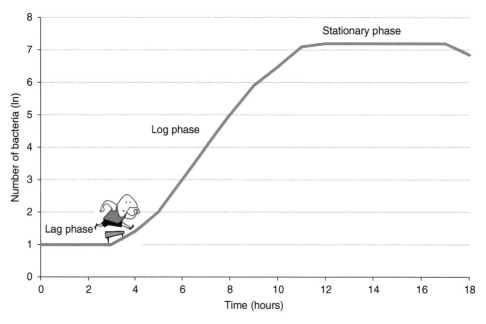

Fig. 2.2. A generic bacterial growth curve depicting time as a hurdle to replication.

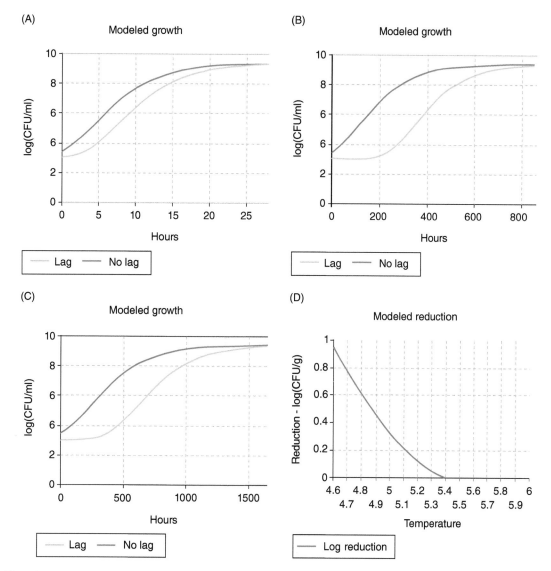

Fig. 2.3. Bacterial growth and reduction curves depicting *Escherichia coli* O157:H7 (USDA, 2003). (A) Under aerobic conditions in broth culture at 25°C, pH 7, 0.5% sodium chloride and initial concentration of 3 colony forming units (CFU)/ml (time is the hurdle). (B) The same conditions except for a temperature of 5°C (time and temperature are hurdles). (C) The same conditions as (B), but with a lower pH of 4.5 (time, temperature and pH as hurdles). (D) Survival during fermentation of Soujouk-style fermented sausage (time, temperature, pH, oxygen content, additives, and competing microflora as hurdles).

corn kernels in a grain truck are post-harvest. Another example would be for livestock, where a live chicken is pre-harvest, and after slaughter is post-harvest.

The environmental conditions of a food system can change during the transition from pre- to post-harvest. The corn kernels in a grain truck have lost the physical protection of the husk and cob, and may have sustained physical damage to the seed coat allowing entry of microorganisms into previously physically protected tissue. The pH and other intrinsic factors of the muscle tissue of a slaughtered animal change post-harvest, and may support rapid growth of inherent and mechanically introduced microbial populations.

Because inoculation sources and intrinsic and extrinsic factors change from pre- to post-harvest, different strategies to control microorganisms need to be considered. Good agricultural practices focus on pre-harvest controls, while good manufacturing practices focus on post-harvest controls.

Gil *et al.* (2015) described preventative pre- and post-harvest measures that lower the risk of food-borne illness from fresh leafy vegetables. They described the production chain challenges to food safety in these main areas:

- Primary production:
  - growing field and adjacent land;
  - animal activity;
  - human activity;
  - climatic conditions of the growing area and climate change;
  - equipment associated with growing and harvesting;
  - manure and soil amendments;
  - water for primary production (water source and irrigation system);
  - production systems; and
  - pre-harvest pathogen testing.
- Post-harvest handling
  - primary preparation;
  - storage and transportation from the field to the packing facility;
  - handling prior processing; and
  - storage and distribution.
- Processing
  - reception and inspection of raw materials;
  - size reduction;
  - washing after cutting;
  - dewatering;
  - weighing and packaging; and
  - storage and distribution.
- Retail food service operation and consumer handling.

Preventative measures are important at all stages of the food production chain, and emphasis was given to the importance of training and educating growers and handlers along the entire continuum. Gil *et al.* (2015) concluded with:

> Thus, dealing with the food safety issues associated with fresh produce it is clear that a "farm to fork" approach is required taking into account a multidisciplinary strategy. Managerial interventions of experts from the food chain including agronomists, food microbiologist, and food science experts are needed.

Summary

Food is not sterile and carries unique microbial ecosystems also known as microbiomes. Growth and survival of microorganisms can be controlled by manipulating environmental factors in food systems. Hurdle technology uses the strategic manipulation of environmental factors to control microorganisms in foods, allowing for better quality food production.

Further Reading

Doyle, M.P. and Beuchat, L.R. (2007) *Food Microbiology: Fundamentals and Frontiers*, 3rd edn. ASM Press, Washington, DC, USA.

ICMSF, International Commission on Microbiological Specifications for Foods (2005) *Microorganisms in Foods 6: Microbial Ecology of Food Commodities*, 2nd edn. Kluwer Academic/Plenum Publishers, New York, USA.

Jay, J.M., Loessner, M.J. and Golden, D.A. (2005) *Modern Food Microbiology*, 7th edn. Springer, New York, USA.

Marchesi, J. (2014) *The Human Microbiota and Microbiome: Advances in Molecular and Cellular Microbiology*. CAB International, Boston, MA, USA.

Shibamoto, T. and Bjeldanes, L. (2009) *Introduction to Food Toxciology*, 2nd edn. Elsevier, New York, USA.

References

Berdy, J. (2005) Bioactive microbial metabolites: a personal view. *Journal of Antibiotics* 58, 1–26.

Bermudez-Aguirre, D. and Corradini, M.G. (2012) Inactivation kinetics of *Salmonella* spp. under thermal and emerging treatments: a review. *Food Research International* 45, 700–712.

Castro-Rosas, J., Gomez-Adapa, C.A., Acevedo-Sandoval, O.A., Ramirez, C.A.G., Villagomez-Ibarra, J.R., Hernandes, N.C., Villarruel-Lopez, A., and Torres-Vitela, R. (2011) Frequency and behavior of *Salmonella* and *Escherichia coli* on whole and sliced jalapeño and serrano peppers. *Journal of Food Protection* 74, 874–881.

Demain, A.L. (1998) Induction of microbial secondary metabolism. *International Microbiology* 1, 259–264.

FDA (2012) *Bad Bug Book: Foodborne Pathogenic Microorganisms and Natural Toxins*, 2nd edn. United Stated Food and Drug Administration, Washington, DC, USA.

Gil, M.I., Selma, M.V., Suslow, T., Jacxsens, L., Uyttendaele, M., and Allende, A. (2015) Pre- and post-harvest preventive measures and intervention strategies to control microbial food safety hazards of fresh leafy vegetables. *Critical Reviews in Food Science and Nutrition* 55, 453–468.

Leistner, L. and Gould, G.W. (2002) Chapter 2: the hurdle concept. In: Leistner, L. and Could, G.W. (eds) *Hurdle Technologies*. Kluwer Academic/Plenum Publishers, New York, USA, p. 17–33.

Liao, C.-H., Cooke, P.H., and Niemira, B.A. (2010) Localization, growth, and inactivation of *Salmonella* Saintpaul on Jalapeño Peppers. *Journal of Food Science* 75, M377–M382.

MMWR (2008) Outbreak of *Salmonella* serotype Saintpaul infections associated with multiple raw produce items – United States, 2008. *MMWR* 57, 929–934.

NRAEF (2014) *ServSafe CourseBook*, 6th edn. Prentice Hall, Upper Saddle River, NJ, USA.

Pan, W. and Schaffner, D.W. (2010) Modeling the growth of *Salmonella* in cut red round tomatoes as a function of temperature. *Journal of Food Protection* 73, 1502–1505.

Pao, S., Long, W., Kim, C., and Rafie, A.R. (2012) *Salmonella* population rebound and its prevention on spray washed and non-washed jalapeño peppers and roma tomatoes in humid storage. *Foodborne Pathogens and Disease* 9, 361–366.

Pin, C., Metris, A., and Baranyi, J. (2013) Next generation of predictive models. In: *Advances in Microbial Food Safety*, Vol. 1. Woodhead Publishing, Cambridge, UK, pp. 498–515.

Remus-Emsermann, M.N.P. and Vorholt, J.A. (2014) Complexities of microbial life on leaf surfaces. *Microbe* 9, 448–452.

Swinnen, I.A.M., Bernaerts, K., Dens, E.J.J., Geeraerd, A.H., and Van Impe, J.F. (2004) Predictive modelling of the microbial lag phase: a review. *International Journal of Food Microbiology* 94, 137–159.

USDA (2003) *USDA-ARS Pathogen Modeling Program, Version 7.0*. United States Department of Agriculture, Agricultural Research Service, Eastern Regional Research Center, Wyndmoor, PA, USA.

3 Intrinsic and Extrinsic Factors and Potentially Hazardous Foods

> **Key Questions**
> - What are intrinsic factors of food that affect microorganisms?
> - What are extrinsic factors of food that affect microorganisms?
> - What is the temperature danger zone?
> - What is a potentially hazardous food?

Environmental Factors that Affect Microorganisms in Foods

In Chapter 2, we introduced the concepts of ecology, environmental factors in foods, and the hurdle concept. In this chapter, we will go into more detail about environmental factors and provide some examples. The environmental factors are categorized as intrinsic and extrinsic (IFT, 2001; Jay *et al.*, 2005; Montville and Matthews, 2007). These concepts will be very important for understanding how specific foodborne pathogens can be problems in foods and food supply networks, and how risk of foodborne illness can be controlled.

According to the United States Food and Drug Administration (FDA):

> Most authorities are likely to divide foods among three categories based on an evaluation of the factors described below: Those that do not need time/temperature control for protection of consumer safety; those that need time/temperature control; and those where the exact status is questionable. In the case of questionable products, further scientific evidence – such as modeling of microbial growth or death, and actual microbiological challenge studies – may help to inform the decision (FDA, 2001).

This categorization of foods builds on the understanding of microbial growth patterns and modeling (Chapter 2) and will be further discussed later in this chapter. First, to provide the proper context, let's categorize and describe the intrinsic and extrinsic factors.

Intrinsic Factors

Intrinsic factors are inherent (or internal) to the food. These result from the chemical, physical, and biological make-up of the food and the resulting interactions of internal environmental factors.

Moisture content and water activity

All biological organisms require water for biological functions and survival. Drying foods is a practice of food preservation that dates back to ancient times. All foods, even dried foods, contain some amount of water. The term for the amount of water in a food is moisture content (MC), which is a numerical value indicating the percentage of water in a food. Not all water in a food item is in a form that would be available to microorganisms to access. Some amount of water remains physically bound to food components tightly enough that it cannot be utilized. The water that is available for microorganisms has the technical term of water activity (a_w). Water activity is defined by Jay *et al.* (2005) as:

$$a_w = p/p_o$$

Where p is the vapor pressure of the solution (the food system) and p_o is the vapor pressure of the solvent (water). The a_w of pure water is 1.00. Addition of solutes lowers the a_w as they bind a portion of the water. Table 3.1 lists examples of foods and their approximate a_w and MC. Note how a_w

and MC are not proportionate, further demonstrating that the amount of hygroscopic solutes in the foods impacts the amount of water available for microorganisms.

Water activity is related to relative humidity (RH) for which:

$$RH = 100 \times a_w$$

An increase in humidity of an environment in which a food is stored can raise the water activity of a food exposed to the air. This has implications for storage of foods, especially dried foods. Dry or low-moisture foods often readily absorb water from the air through hygroscopy. Packaging and storage conditions for such foods is important for quality and microbial food safety.

Table 3.2 lists examples of pathogenic bacteria and the approximate a_w at which their growth can be supported. Most bacteria require an a_w value of at least 0.9 to grow. An important foodborne pathogenic bacterial species to note as an exception is *Staphylococcus aureus*, which will be further described in Chapter 6. Most fungi (molds and yeasts) can grow at a_w at or above 0.8. Xerophyllic (dry-loving) fungi can grow at a_w as low as 0.61. The term "alarm water" is used for dried foods to indicate the water content that should not be exceeded if mold growth is to be avoided (Jay *et al.*, 2005). Low moisture as a control for microbial growth, in itself, is not always an effective way to prevent foodborne disease (see Box 3.1).

pH and acidity

Hydrogen ion concentration (pH) is a critical factor affecting microorganisms in foods. The more acidic the food (higher acidity), the lower the pH value. Most microorganisms have an optimal pH around a value of 7, a neutral pH that is neither acidic nor basic. At pH values lower than 7, the condition becomes increasingly acidic. At pH values higher than 7, the condition becomes basic. Changes in the pH affect enzyme functionality, cellular homeostasis, and chemical reactions in the food environment. At a neutral pH, the microorganism does not have to expend as much energy on maintenance of an internal homeostatic pH, freeing resources for growth. Table 3.2 lists the minimum, optimal, and maximum pH levels at which growth has been shown to occur for the pathogenic microorganisms listed. Growth rates at the minimum and maximum levels tend to be for adapted strains and are much slower than at optimal pH. See Box 3.2 for the example of acid-adapted *Salmonella*.

Table 3.1. Approximate water activities (a_w) and moisture contents (MC) for selected food examples.

Foods	a_w	MC (%)
Fresh meat, poultry, fish	0.91->0.99	63.2–79.1
Eggs	0.97	75.3
Fresh fruits and vegetables	0.97–1.00	79.0–96.0
Bread	0.94–0.97	36.0
Cured meat	0.87–0.95	23.0–60.0
Cake	0.90–0.94	9.0
Cake icing	0.76–0.84	13.5
Flour	0.65–0.87	10.3–11.0
Honey	0.75	17.0
Sugar	0.19	0.07
Crackers	0.10	6.0

Adapted from AquaLab (2015), IFT (2001), FoodTechSource (2002), and Manitoba Agriculture (2015).

Table 3.2. Minimum, optimal and maximum pH values and minimum water activities (a_w) at which example microorganisms will grow.

Microorganism	Minimum pH	Optimal pH	Maximum pH	Minimum a_w for growth
Salmonella spp.	4.0	7.0	9.0	0.94
Listeria monocytogenes	4.1	6.0–8.0	9.6	0.92
Escherichia coli	4.5	7.0	9.0	0.95
Clostridium botulinum	4.7	7.0	8.3	0.93
Bacillus cereus	5.0	7.0	9.5	0.93
Shigella spp.	5.0	7.0	9.2	0.97
Vibrio parahemolyticus	5.0	7.0	11.0	0.94
Staphylococcus aureus	4.4	7.0	9.4	0.86
Certain molds	0.3	7.0	11.0	0.61

Adapted from Jay *et al.* (2005) and IFT (2001).

Box 3.1. The example of *Salmonella* in low-moisture foods

Many enteric bacteria do not survive in dry environments. Some do possess the ability to survive in a dormant state, and these include *Salmonella*. Most documented outbreaks linked to low-moisture foods have been caused by *Salmonella*. Examples of low-moisture foods linked to *Salmonella* outbreaks include powdered milk, powdered infant formula, chocolate, potato chips, peanut butter, cereal, almonds, peanuts, tea, pepper, dog food, and tahini paste (Finn *et al.*, 2013).

Bacteria can gain protection from moisture loss within food matrix components such as fat and other solids. Other intrinsic and extrinsic factors such as redox potential, temperature and modified atmosphere packaging can increase the chances for surviving under low-moisture conditions. In fact, microbiologists use similar conditions for long-term storage of bacterial cultures through the process of lyophilization. Lyophilization is the process of freeze drying, and microbial cultures are often processed this way in milk for the protective effect of the matrix materials.

The presence of *Salmonella* can cause serious food safety problems in low-moisture foods as the traits involved with low-moisture survival also relate to increased pathogenicity or ability to cause disease, and increased resistance to heating. Therefore, their presence in a food is a high risk for foodborne illness, and control of growth is not sufficient to control the risk. Sanitation, pest control, and testing (keeping enteric bacteria out of the food environment) are more practical controls for this risk.

Box 3.2. The example of acid-adapted *Salmonella*

Salmonella are environmentally adaptable bacteria, a trait that challenges food safety control strategies. This type of adaptability allows bacteria to respond to environmental stress due to changing conditions. The acid tolerance response (ATR) is a phenomenon whereby bacteria are able to adapt to lowering of pH. The mechanisms by which adaptation occurs can also enhance the ability of bacteria to be pathogenic.

Alvarez-Ordonez *et al.* (2010) evaluated the ATR of *Salmonella typhimurium* subjected to culture pH adjustments with different acids (organic acids and hydrochloric acid) at different temperatures. They determined that temperature was a critical factor in conjunction with pH for the impact of ATR on growth of *Salmonella*. In a temperature range of 25–37°C, the minimum pH that supported growth was 4.5 using citric and hydrochloric acids, and 5.4 and 6.4 for lactic and acetic acids, respectively. Colder temperatures resulted in a decrease in ATR, whereas warmer temperatures up to 45°C enhanced ATR. Temperature control was reinforced as an important control for risk of *Salmonella* growth.

The FDA defines low-acid foods and acidified foods (CFR, 2000; McGlynn, 2010). Foods at pH levels lower than 4.6 are considered high-acid foods and at pH values of 4.6 or higher low-acid foods. Low-acid foods, or foods with pH values of 4.6–8.0, are more likely to support the growth of pathogenic bacteria. This has implications for how foods can be processed and stored, especially for canned or hermetically sealed products. Table 3.3 lists approximate pH values for some example foods. Although growth rates are slowed for microorganisms as pH is lowered, it is not always an assurance that an adapted pathogen won't emerge as a hazard (see Box 3.2).

Nutrient content

All microorganisms require food sources to survive and grow. Some microbes require a complex mixture with many growth factors, i.e. most Gram-positive bacteria. Gram-negative bacteria can synthesize growth factors from a more limited array of nutrients. Some fungi are able to grow in distilled water due to their mechanisms for scavenging limited nutrients.

Because food is food, any food used by humans will also be a nutrient source for microorganisms. Foods higher in proteins and carbohydrates tend to be the richest sources of energy and building materials for both humans and microorganisms.

Table 3.3. Approximate pH values for example foods.

Food	pH
Egg white	7.0–9.0
Egg yolk	6.4
Camembert cheese	7.4
Cheddar cheese	5.9
Milk	6.3–8.5
Buttermilk	4.5
Fresh fish	6.6–6.8
Ground beef	5.1–6.2
Bread	4.4–4.5
Potatoes	5.7–6.1
Spinach	5.5–6.8
Squash	5.5–6.2
Tomato	4.2–4.9
Cabbage	5.2–6.9
Sauerkraut	3.4–3.6
Bananas	4.5–5.2
Raisins	3.8–4.0
Strawberries	3.0–3.5
Lemons	2.2–2.4
Vinegar	2.0–3.4

Adapted from CFSAN (2014).

Redox potential

All microorganisms metabolize food for energy. Strict aerobic microorganisms must have oxygen for this process. Some microorganisms have multiple mechanisms for metabolism and can use oxygen or go without it; these are called facultative aerobes (or facultative anaerobes). For strict anaerobic bacteria, oxygen can be toxic, and they will only grow in the absence of oxygen. Some bacteria are described as microaerophilic and require small amounts of oxygen. All fungi are aerobic, but some have mechanisms for scavenging oxygen that allow them to grow in low-oxygen environments. Most pathogenic bacteria are facultative aerobes, so control of oxygen content alone is not an effective method of risk control.

The redox potential (E_h) of a food is the ratio of the oxidizing power (electron accepting) to the total reducing power (electron reducing). The units to report E_h values are millivolts (mV). Negative values indicate a low E_h, which is indicative of anaerobic conditions. Positive values are more aerobic.

Antimicrobial chemicals

All food is made of chemicals. Chemicals are naturally intrinsic to food, mixed in as additives, and result from chemical reactions in the food. Some of these have toxic chemical effects that lead to antimicrobial activity. All chemicals have potential for toxicity depending on dosage and the affected organism. So-called natural chemicals are not necessarily different in terms of toxicity in comparison to manufactured chemicals, a common misperception among consumers. Antimicrobial agents used in foods affect select target organisms and have a very low risk of harm to humans when used at proper concentrations.

Crozier-Dodson *et al.* (2005) classified antimicrobial ingredients into six categories:

- Acid antimicrobial – This category includes organic acids (i.e. acetic, lactic, propionic), salts of lactic acid (lactates), buffered sodium citrate, and acidified sodium chlorate. Synergistic effects can be achieved for some of these when combined with other antimicrobial agents. Solubility and pH affect the antimicrobial strength of organic acids.
- Chemical antimicrobial – This category includes trisodium phosphate, chlorine dioxide, peracetic acid, sodium nitrate, and nitrate. These have specified uses at regulated concentrations, as do most antimicrobial ingredients.
- Ovo-antimicrobial – Lysozyme, an enzyme intrinsic to animals and plants, has antimicrobial activity and is likely a product of evolution for natural defenses against infections. Chicken egg albumin is used as a source for commercial lysozyme and is used in processed meats.
- Lacto-antimicrobial – Lactoferrin, a glycoprotein in milk, chelates (binds) iron, which can have antimicrobial properties of nutrient depletion for certain types of microorganisms. It is used in food animal carcass treatments to help limit bacterial attachment.
- Bacto-antimicrobial – Bacteriocins are small peptides produced by certain bacteria. These have antimicrobial activity against other types of bacteria, including some important pathogens. These give the producing bacteria a biological advantage, and can be manipulated to be used as a chemical additive. Nisin is a commercially available bacteriocin approved by the FDA for use in meat and dairy products.
- Phyto-antimicrobials – Many plants produce antimicrobial compounds as defense mechanisms. Some of these can be used for food preservation

and safety. Spices and herbs commonly contain chemicals described as essential oils, and these often have antimicrobial activity. However, the strong flavors and odors of essential oils can limit their practical application. Other examples of plant extracts that can be utilized as antimicrobial ingredients are tannins and hop acids.

Competitive microflora

Food is not sterile, and otherwise harmless bacteria can be formidable competition for pathogens. Some produce bacteriocins and organic acids along with other antimicrobial chemicals described in the previous section. Competitors can grow quickly and outstrip nutrients or build up waste products that affect other intrinsic factors such as pH and E_h. Fermented foods are good examples of how microbes can help preserve food through competitive exclusion. Processing a food and reducing the native microorganisms can make the food more vulnerable to post-processing contamination and growth of certain pathogens.

Biological structure

Foods with natural barriers lower the risk of foodborne illness due to an innate ability to physically prevent microorganisms from reaching nutrient-rich internal environments. Examples of these natural barriers include shells (i.e. egg shells), husks (i.e. nutshells), skins (i.e. animal skins), and rinds and peels. Damage to these structures tends to lead to rapid microbial spoilage in fresh foods.

Homogeneity

This is a trait of foods that is often overlooked. Foods and ingredients are often mixed together. These mixtures can range in terms of level of homogeneity. Homogenous mixtures are those that are thoroughly mixed and have the same composition throughout. The example of a lettuce salad with other vegetables and foods mixed in is a much less homogenous mixture than the salad dressing. In less homogenous mixtures, there can be differences in intrinsic factors of microenvironments in different areas of the mixture.

Consider generic American potato salad. The basic ingredients are cooked potatoes, chopped onions, and mayonnaise (many variations of this exist). Consider what the riskiest ingredient is for

supporting growth of microorganisms. Often the mayonnaise is blamed; however, it is a homogenous mixture of oil, emulsifier, and vinegar. The pH of mayonnaise is very low and the acetic acid of the vinegar has antimicrobial activity beyond the effect of lowering pH. In contrast, the pieces of cooked potato are very welcoming environments for microorganisms to grow. The pH is favorable, there is energy-rich starch, and the competitors have been knocked down by cooking. The mayonnaise actually has a preservative effect if it is mixed in well enough to penetrate the tissues of the other ingredients.

Another example is the differences between grain and milled grain. These differ in particle size and the level of homogeneity. Think of grain stored in a large grain bin. There are many microenvironments and microecologies in such a facility. Hot spots are pockets of biological activity induced by moisture pockets, insects, mites, fungi, and other microorganisms, where the heat generated can actually char the grain (Sinha and Wallace, 1965). These are not evenly distributed throughout the mass of stored grain. Similar activity is less likely in bulk storage of milled grain products due to the moisture content being more homogenously distributed and less likely to support such microenvironments.

Extrinsic Factors

Extrinsic factors are environmental factors external to the food. Although external, these factors can impact the intrinsic factors of the food, thereby affecting the biological activity of microorganisms in the food.

Packaging and atmosphere

The packaging that food is stored in affects the type of atmosphere in contact with the food. If the package is porous and allows air exchange, then the food is exposed to more atmospheric oxygen. Changes to the atmosphere surrounding a food can affect the E_h and pH of the food, especially at the surface of the food. If the package is sealed, then there can be more control over the types of gas concentrations in the atmosphere surrounding the food. This approach is called modified atmosphere packaging. See IFT (2001) and Phillips (1996) for a more in-depth explanation of how controlling concentrations of gases in packaging can have inhibitory effects on certain microorganisms in foods. It is important to understand that there are limitations to this form of hazard control. Controlling the atmosphere to slow spoilage of microorganisms can increase the risk presented by some microbial pathogens including *Listeria monocytogenese* and *Clostridium botulinum* (see Chapter 6).

Time

Time is considered an extrinsic factor. This relates to the length of time that a food is kept under particular conditions. In Chapter 2 (see Figure 2.2), time was an initial hurdle to overcome for growth of bacteria. Many food products are labeled with dates for best use by. This is often related to estimates of time for growth of spoilage or pathogenic microorganisms to initiate. "When time alone is used as a control, the duration should be equal to or less than the lag phase of the pathogen(s) of concern in the product in question" (IFT, 2001). The lag phase can be lengthened through combinations of controls of intrinsic and extrinsic factors, as with hurdle technology (see Chapter 2). Time is an increasingly important factor to consider with the globalization of the world's food supply and the time it takes to move the food around.

Temperature

Temperature is a critical factor for controlling microbial growth as it impacts the enzymes and other biochemical reactions required for survival. The biological diversity of microorganisms is immense, and because of this, microorganisms are categorized into five groups:

- Psychrophiles (cold loving) can grow below –20°C, with an optimum temperature range of –20 to 0°C. These are environmental extremophiles and are not typically associated with foods.
- Psychrotrophs can grow at temperatures below 7°C with an optimum temperature range of 20–30°C. This means they can grow at refrigeration temperatures, but will grow slowly (longer lag phase at lower temperatures). Few pathogens are included in this category, but some important ones are, such as *L. monocytogenes* (see Chapter 6).
- Mesophiles (warm loving) can grow in the temperature range of 20–45°C, with an optimum temperature range of 30–40°C. Most enteric foodborne pathogens fall into this category.

- Thermophiles (hot loving) grow only at temperatures higher than 45°C. Thermophilic bacteria associated with foods have optimum temperature ranges for growth between 55 and 65°C. Other environmental extremophiles not typically associated with foods have been isolated from environments as hot as 100°C.
- Thermoduric microorganisms are mesophiles that can survive periods of time at high temperatures. These are spore-forming microorganisms that in their resistant spore form can survive heat exposures such as cooking or canning. After the heat exposure, and once temperatures drop into the mesophilic range, these microbes tend to have very fast growth rates (steep curves for the logarithmic portion of the growth curve). Bacterial and fungal spoilage microorganisms are important problems in this category, as well as certain pathogens including *Bacillus* and *Clostridium* (see Chapter 6) species.

In food safety, the temperature danger zone is used as a way to describe the storage temperatures of highest risk for growth of pathogenic microorganisms. The temperature danger zone is the temperature range of 4–60°C (40–140°F) (FSIS, 2013; see Fig. 3.1). Pyschrotrophs and mesophiles can grow rapidly within this range of temperatures. Management of foods that are held within this temperature range includes the critical control of time. In general, potentially hazardous foods (PHFs, see description below) should be limited to 2 hours within the temperature danger zone. Potentially hazardous foods that need to be out for longer periods of time should be held at over 60°C (140°F) or below 4°C (40°F) – keep hot foods hot and cold foods cold. Temperature abuse is a term used for keeping potentially hazardous foods in the danger zone for too long, allowing for the possibility of growth of pathogenic microorganisms.

Kim *et al.* (2013) did a study of how the internal temperature of foods was affected for select foods stored in a car trunk exposed to sunlight with varying cloud cover. Temperatures were measured every 10 minutes for 3 hours. The initial and final temperatures of the foods tested under the warmest conditions are summarized Table 3.4. This study demonstrates how consumer practices for transportation of food from retailer to home can affect food temperature. The authors encouraged consumers to use coolers and ice packs when long transportation times are needed. Readers are referred to Kim *et al.* (2013) for a more in-depth analysis.

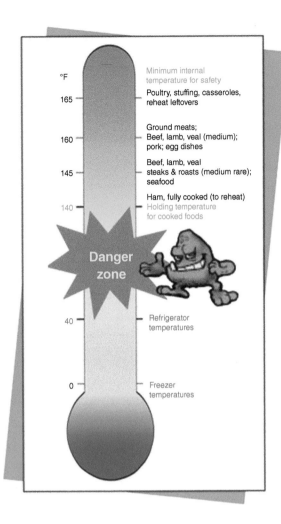

°F

165 — Poultry, stuffing, casseroles, reheat leftovers

Minimum internal temperature for safety

160 — Ground meats; Beef, lamb, veal (medium); pork; egg dishes

145 — Beef, lamb, veal steaks & roasts (medium rare); seafood

140 — Ham, fully cooked (to reheat) Holding temperature for cooked foods

Danger zone

40 — Refrigerator temperatures

0 — Freezer temperatures

Fig. 3.1. A graphic produced by the United States Department of Agriculture Food Safety and Inspection Service Fight Bac campaign (USDA, 2015) used for educating consumers about the temperature danger zone.

Table 3.4. Change of internal temperature of food items left in a warm car trunk.

Item	Initial temperature (°C)	Temperature after 3 hours (°C)
Car trunk air	31.7 ± 0.7	37.6 ± 1.3
Egg	4.6 ± 0.9	33.2 ± 2.0
Milk	4.8 ± 0.9	32.1 ± 0.8
Tofu	4.6 ± 0.8	32.9 ± 0.6
Fresh meat	4.6 ± 0.6	34.4 ± 0.6
Frozen meat	−15.8 ± 0.9	30.5 ± 2.0

Data are from Kim *et al*. (2013).

Temperature abuse can happen at many points in food supply networks. Multiple points of temperature abuse can have cumulative effects on biological activity. Monitoring the time and temperature throughout the food supply network is an important risk control measure. See Box 3.3 for an example of pathogen risk being higher with complexity in bringing peppers to the consumer.

Relative humidity

Humidity was mentioned earlier as an environmental factor that can affect MC and a_w. It is an external factor because it is the moisture content of the air surrounding the food. It is important to control humidity in areas where dried foods are stored and during transportation. Dry foods tend to be hygroscopic and will readily pick up moisture from the air.

Potentially Hazardous Foods

Potentially hazardous foods (PHFs) are basically foods for which the intrinsic and extrinsic factors would

Fresh produce is necessary for a healthy diet. More than ever before, fresh produce is traveling around the world to reach wider markets. In the past 10 years, more fresh produce has been brought to consumers in ways to increase convenience, such as pre-washed and cut (Wang and Ryser, 2014). Although washing can help reduce risk, there are increased opportunities for cross-contamination and temperature abuse of cut or newly damaged plant tissues. These opportunities for contact and release of nutrients can increase the risk of contamination with pathogens, such as *Salmonella*, causing widespread outbreaks at the end of food supply networks.

The salmonellosis outbreaks described in Chapter 7 (Box 7.3) are examples of how large-scale outbreaks linked to produce can happen.

support the growth of pathogenic microorganisms. The FDA has provided a regulatory definition for a PHF. This term is no longer used and has since changed to "time/temperature control for safety food" (TCS) based on recommendations from the study done by the IFT (2001). Jol *et al.* (2006) provide an explanation of the history of the terminology used for categorizing foods based on food safety risk. The definition for TCS is based on the need for time/temperature control to control for risk of foodborne illness and applies to particular foods. Consideration is given to interactions of pH and a_w (FDA, 2013).

Foods that fall into this category are subject to regulatory rules, which can have broad implications. Care is needed in how to categorize foods to ensure safety and limit waste.

> Scientifically sound criteria for determining whether foods require time/temperature control for safety should consider 1) processes that destroy vegetative cells but not spores (when product formulation is capable of inhibiting spore germination; 2) post-process handling and packaging conditions that prevent reintroduction of vegetative pathogens onto or into the product before packaging; and 3) the use of packaging materials that while they do not provide a hermetic seal, do prevent reintroduction of vegetative pathogens into the product (IFT, 2001).

Pathogen environmental monitoring (PEM) programs are an important aspect of preventative food safety strategies. Frier and Shebuski (2014) describe the aspects of PEM programs, including management commitment, determination of need for PEM programs, risk evaluation, sampling plan, sampling method and evaluation of results. A PEM program is a prerequisite to hazard analysis and critical control points, which is explained in Chapter 10.

Summary

Intrinsic and extrinsic factors are the environmental conditions that impact growth and survival of microorganisms in foods. Food safety experts need to have a good understanding of these factors and how they work together. Control of single factors is not typically adequate for effective control, and the overall combination of factors determines whether a food is potentially hazardous.

Further Reading

Doyle, M.P. and Beuchat, L.R. (2007) *Food Microbiology: Fundamentals and Frontiers*, 3rd edn. ASM Press, Washington, DC, USA.

Jay, J.M., Loessner, M.J., and Golden, D.A. (2005) *Modern Food Microbiology*, 7th edn. Springer, New York, USA.

References

Alvarez-Ordonez, A., Fernandez, A., Bernardo, A., and Lopez, M. (2010) Acid tolerance in *Salmonella typhimurium* induced by culturing in the presence of organic acids at different growth temperatures. *Food Microbiology* 27, 44–49.

AquaLab (2015) Intro to water activity. Available at: http://www.aqualab.com/education/intro-to-water-activity/ (accessed 5 January 2015).

CFR (2000) *Code of Federal Regulations*, Title 21 part 114. Office of the Federal Register, National Archives and Records Administration, Washington, DC, USA.

CFSAN (2014) pH values of various foods. Available at: http://www.fda.gov/Food/FoodborneIllnessContaminants/CausesOfIllnessBadBugBook/ucm122561.htm (accessed 5 January 2015).

Crozier-Dodson, B.A., Carter, M. and Zheng, Z. (2005) Formulating food safety: an overview of antimicrobial ingredients. *Food Safety Magazine* January. Available at: http://www.foodsafetymagazine.com/magazine-archive1/december-2004january-2005/formulating-food-safety-an-overview-of-antimicrobial-ingredients/ (accessed 22 December 2014).

FDA (2001) Evaluation and definition of potentially hazardous foods. Available at: http://www.fda.gov/Food/FoodScienceResearch/SafePracticesforFood Processes/ucm094145.htm. (accessed 5 October 2016).

FDA (2013) Food code. Food and Drug Administration, United States Department of Health and Human Services. College Park, MD, USA. Available at: http://www.fda.gov/Food/GuidanceRegulation/Retail FoodProtection/FoodCode/ucm374275.htm (accessed 22 December 2014).

Finn, S., Condell, O., McClure, P., Amezquita, A., and Fanning S. (2013) Mechanisms of survival, responses and sources of *Salmonella* in low-moisture environments. *Frontiers in Microbiology* 14, 1–15.

FoodTechSource (2002) Determining moisture content in foods. Available at: http://www.foodtechsource.com/rcenter/tech_data/td_moisture.htm (accessed 21 January 2015).

Frier, T. and Shebuski, J. (2014). Components for an effective pathogen environmental monitoring program. *Food Quality and Safety* February/March. Available at: http://www.foodquality.com/details/article/5862961/Components_for_an_Effective_Pathogen_Environmental_Monitoring_Program.html?tzcheck=1 (accessed 30 January 2015).

FSIS (2013) "Danger Zone" (40 °F – 140 °F). Available at: http://www.fsis.usda.gov/wps/portal/fsis/topics/food-safety-education/get-answers/food-safety-fact-sheets/safe-food-handling/danger-zone-40-f-140-f/ct_index (accessed 21 January 2015).

IFT (2001) Evaluation & definition of potentially hazardous foods. A Report of the Institute of Food Technologists for the Food and Drug Administration of the United States Department of Health and Human Services, Task Order No. 4. Washington, DC, USA. Available at: http://www.fda.gov/Food/FoodScienceResearch/SafePracticesforFoodProcesses/ucm094141.htm (accessed 22 December 2014).

Jay, J.M., Loessner, M.J., and Golden, D.A. (2005) Chapter 3: intrinsic and extrinsic parameters of foods that affect microbial growth. In: Jay, J.M., Loessner, M.J., and Golden, D.A. (eds) *Modern Food Microbiology*, 7th edn. Springer, New York, USA, pp. 39–59.

Jol, S., Kassianenkio, A., Wszol, K., and Oggel, J. (2006) Issues in time and temperature abuse of refrigerated foods. *Food Safety Magazine* December/January. Available at: http://www.foodsafetymagazine.com/magazine-archive1/december-2005january-2006/issues-in-time-and-temperature-abuse-of-refrigerated-foods/ (accessed 21 January 2015).

Kim, S.A., Yun, S.J., Lee, S.H., Hwang, I.G., and Rhee, M.S. (2013) Temperature increase of foods in car trunk and the potential hazard for microbial growth. *Food Control* 29, 66–70.

Manitoba Agriculture (2015) Water content and water activity: two factors that affect food safety. Available at: http://www.gov.mb.ca/agriculture/food-safety/at-the-food-processor/water-content-water-activity.html (accessed 21 January 2015).

McGlynn, W. (2010) *The Importance of Food pH in Commercial Canning Operations. Food Technology Fact Sheet*. Robert M. Kerr Food & Agricultural Products Center, Oklahoma State University, Stillwater, OK, USA.

Montville, T.J. and Matthews, K.R. (2007) Chapter 1: Growth, survival, and death of microbes in foods. In: Doyle, M.P. and Beuchat, L.R. (eds) *Food Microbiology: Fundamentals and Frontiers*, 3rd edn. ASM Press, Washington, DC, USA, pp. 3–22.

Phillips, C.A. (1996) Review: modified atmosphere packaging and its effects on the microbiological quality and safety of produce. *International Journal of Food Science and Technology* 31, 463–479.

Sinha, R.N. and Wallace, H.A.H. (1965) Ecology of a fungus-induced hot spot in stored grain. *Canadian Journal of Plant Science* 45, 48–59.

USDA (2015) FSIS Educational Campaigns. Available at: http://www.fsis.usda.gov/wps/portal/fsis/topics/food-safety-education/teach-others/fsis-educational-campaigns (accessed 10 June 2015).

Wang, H. and Ryser, E.T. (2014) Microbiological safety of fresh-cut produce from processor to your plate: assessing the key steps to managing microbial risks to produce from farm to fork. *Food Safety Magazine* August/September, 64–70.

4 Humans and Microbes – Risk Analysis

Key Questions

- What is risk analysis?
- How do microbial hazards become risks to humans?
- How do lapses in policies result in type I and type II food risks?
- How do public–private synergies impact efficient policy design?

Introduction to the Risk Analysis Model

In Chapter 1, we provided a brief description of the food supply network complexity, and in Chapter 3, we discussed how intrinsic and extrinsic factors in these networks would support the growth of pathogenic microorganisms and the concept of potentially hazardous foods. In this chapter, we introduce risk analysis, which is used to examine why food safety problems can occur and how to control food risk. Hazardous foods must be consumed before food safety problems occur. This chapter also answers the question of how microbial hazards are introduced to foods and how they become a risk to humans. Lapses in the food supply network are used to explain this phenomenon in more detail.

Figure 4.1 shows a schematic presentation of the risk analysis model. It is a model that we use to discuss how components of the system interact to generate food safety risks. It has three major components (risk assessment, risk management, and risk communication), all interacting to assist policy makers in designing science-based and coordinated regulatory systems and standards. The risk analysis model provides a framework to understand food risks and to assist in cost-effective risk mitigation strategies (WHO, 2015). Gaps in information sharing impact a unified strategy in minimizing microbial food safety risks.

Risk assessment and responsibilities for food safety risks

Risk assessment addresses the question of "what can go wrong." In our context, the question will be used to determine "what is the probability that hazardous foods will become food safety risks to humans?" and "what are the economic consequences when this occurs?" It is the process of identifying a hazard and evaluating the risk of a specific hazard, whether in absolute or relative terms. Another definition is "the scientific evaluation of known or potential adverse health effects resulting from human exposure to foodborne hazards" (WHO, 2016).

A major requirement of risk assessment is that it should be comprehensive, feasible (given time and resources), and directed toward assisting those responsible for making policy decisions. The risk assessment steps involve:

1. Hazard identification: The element or elements that pose potential harm, or an act or phenomenon that has the potential to produce hazardous foods or other adverse consequences to humans.
2. Hazard assessment: A description of what might go wrong and how it might happen. We have to determine if exposure to an agent causes an increased incidence of harm.
3. Dose–response: For a given dose, what response does an environment or a body give? Characterize the relationship between hazard (at different levels

along the supply chain and doses consumed) and the incidence of the adverse health effect.

4. Exposure analysis: How does exposure occur, where does exposure occur, and for how long does it occur? Measure or estimate the intensity, frequency, and duration of actual or hypothetical exposure of humans to the identified food risk agent (e.g. *Salmonella*).

5. Economic outcome assessment and risk measurement: What is the magnitude of the outcome if a food risk occurs? In other words, what are the economic consequences if a hazard occurs in terms of loss of business and revenue, cost of illness and death, and how many and who are affected? We measure risk by estimating the probability and magnitude of specific harm to an exposed individual or population based on information from steps 1 to 4.

Figure 4.2 shows a summary of the risk assessment for a hypothetical hazardous food. It shows how

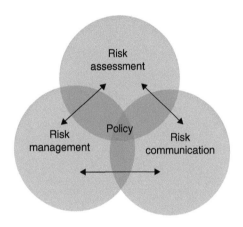

Fig. 4.1. Schematic presentation of the risk analysis model.

the probability of exposure could be estimated for infected cattle. In this example, the probability of importing infected cattle is estimated to be 3.3%.

The risk assessment process has transitioned from the traditional qualitative inspection system to a quantitative process. Quantitative risk assessment models require extensive data acquisition and analysis. In addition, the models must also be dynamic to account for variability that can influence the overall model. A comprehensive risk assessment involves a multidisciplinary team of researchers from agencies, academia, and industry.

Risk management and responsibilities for food safety risks

Risk management is used in determining what action to take to reduce or eliminate food safety risks. The determination is based on information from hazard identification and risk assessment along with information on technical resources, and social, economic, and political values. It also includes the design and implementation of strategies and policies. Risk management could be defined as the process of weighing policy alternatives to accept, minimize, or reduce assessed risks and to select and implement appropriate intervention alternatives (WHO, 2015). Effective risk management should address economic issues related to:

- Would you get more, or greater, benefit if resources were used somewhere else … maybe to address another hazard?
- Can you get the same degree of risk reduction at a lower cost (e.g. increase the efficiency of resource allocation)?
- When does the cost of further risk reduction outweigh the benefits?

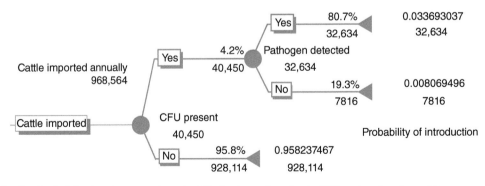

Fig. 4.2. Simple decision tree to assess to probability the hazardous foods. CFU, colony-forming units.

Risk management enables us to:

- avoid or eliminate risk;
- decrease or modify the risky activity;
- decrease the magnitude or frequency of risk; and
- decrease the vulnerability of exposed people or property.

The major issue in the science of risk management is to identify the lowest cost alternative with the greatest impact to reduce risk and to determine cost–risk trade-off, or who is liable to pay for food safety loss. These are complex issues that require public–private partnerships (everyone working together). Lapses in public–private partnership create major implementation conflict and likely morally hazardous behavior. At the core, government or policy makers should assume the responsibility of risk management to ensure efficient resource allocation. The responsibility of the government is needed because food safety is a global good and a minimum level of safety is expected by society (Jensen and Unnevehr, 2009). Strategies to mitigate risks could include insurance schemes, common law, government intervention, and standards/voluntary private sector self-regulation.

Risk communication and responsibilities for food safety risks

Risk communication is the exchange of information between concerned parties about levels of health or environmental risk, the significance and meaning of those risks and the decisions, and actions or policies aimed at managing or controlling the risks. Sandman's axiom characterized risk as a proxy of hazard plus outrage (risk = hazard + outrage). Hazard could be equivalent to hazardous foods and outrage could be defined as "dread" or "fear of the unknown." This specification suggests two corollaries:

1. When hazard is high and outrage is low, scientists are concerned.
2. When hazard is low and outrage is high, the public is concerned.

Factors affecting hazard and outrage lead to the concept of "risk perception." These factors could include catastrophic potential, familiarity with the food risk, understanding, controllability (personal, external from industry and agency policy and

standards), willingness of exposure, effects on children, effects manifestation, effects on future generations, victim identity, dread, trust in institutions, media attention, accident history, equity, benefits, reversibility, and origin.

The United States Environmental Protection Agency's seven rules of risk communication comprise:

1. Accept and involve the public as a legitimate partner.
2. Plan carefully and evaluate your efforts.
3. Listen to the public's specific concerns.
4. Be honest, frank, and open.
5. Coordinate and collaborate with other credible sources.
6. Meet the needs of the media.
7. Speak clearly and with compassion.

Risk communication recommendations include:

- use simple, graphical material;
- provide opportunity for learning;
- put risks into perspective;
- relate on a personal level;
- understand qualitative concerns;
- recognize the impact of subtle changes in problem formulation;
- identify specific target audiences;
- generate involvement;
- avoid high-threat campaigns;
- use multiple channels and media;
- use peer and social relationships; and
- be inventive.

As food supply networks become increasingly technological and connected, risk communication becomes more complex. To understand how risk perceptions are socially constructed through the interaction of consumers, scientists, industry representatives, the media, and food safety opinion leaders, risk communication science needs to draw upon the diffusion of innovations from social science research.

The Interface Model: How Microbial Hazards become a Risk to Humans

In the previous section, we stated that multiple agencies and stakeholders are involved with risk assessment, communication, and management. Policy making is at the core of effective risk mitigation strategies. Results from risk management should be provided to policy makers with sufficient information to formulate effective and efficient policies. However, lapses in microbial food risk are due to several reasons:

- risk communications (e.g. complexity of food supply networks and limitations with information sharing among stakeholders and agencies);
- risk management (e.g. limitations with adequate resources); and
- ineffective policy response, i.e. timely and targeted (e.g. limitations with control and prevention measures).

We use a control-oriented framework to discuss the interface model (Fig. 4.3), showing how lapses in control from communication, management, or policy can lead to vulnerability and increased food risk. The case of orange juice processing is used to identify the source of vulnerabilities that present the possibility of foodborne illness risks to humans. Figure 4.3 displays a conceptual model, which provides a systemic view of investments (risk management) and vulnerabilities (risk assessment) and the relationship between them. How managers respond to the opportunity for controlling food risk and investments governs the evolution of microbial risks to foodborne risks for humans.

Causes of threats and interface of microbial food risks

Figure 4.3 shows that the failure to invest in risk management tools and continuously reduce food-borne illness risks could be a high risk and cost for all participants along the food supply chain. Beginning in the upper left corner of the figure, it seems self-evident that threats have causes. In most security-oriented studies, the cause of threats is treated purely as exogenous. As shown in Fig. 4.3, in a control-oriented theory, threats could be exogenous or endogenous (e.g. directly from a contaminated product or indirectly from cross-contamination). The cause of the threat may initially be poorly understood, but an important goal of the interface model should be to understand the cause in order to eliminate or control it (Bohn, 1994; Lee and Whang, 2005). The improved knowledge of control factors achieved through the diagnostic process often results in preventative measures. The whole purpose of control-oriented food supply chain security systems is to estimate and control threats. This means that a threat must be perceived, since a threat that is not anticipated cannot be estimated, controlled, or defended against (Nganje and Skilton, 2011). Threats that are not controlled increase foodborne illness risks, depicted with a positive (+) risk of defect.

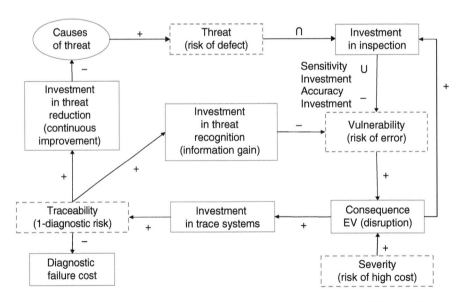

Fig. 4.3. Control oriented food supply network and the interface framework (Nganje and Skilton, 2011).

Investments in threat reduction

The main tenet of microbial food risk management is the investment of limited resources. We assume that as threats increase, participants in food supply chains increase their investment in the system designed to detect food risks before they become foodborne illness outbreaks. In control-oriented food supply chain security systems, investments could relate primarily to inspection systems or control measures such as hazard analysis critical control points. Investments can reduce the vulnerability of humans to food risks, indicated by (−) in Fig. 4.3. Nganje and Skilton (2011) suggest that there are decreasing returns to detection systems, such that as the probability of defects or attacks increases, improvements in detection resulting from additional investments will diminish. When a threat is low, the benefit of additional investments in inspection systems (such as frequent sampling) will be constrained by limited resources. As threats increase, benefits from additional investments will initially rise, and then plateau. Highly probable food risks will be easier to detect with lower sampling rates and lower levels of investment, so that the form of the relationship between threat and investment in detection is likely to take an inverted U shape (Fig. 4.3).

There are multiple avenues to invest in the interface model and reduce the vulnerability of microbial food risks to humans. These include:

- investments in traceability (this will be expanded upon in Chapter 14);
- investments in technology and information (e.g. hazard analysis and critical control points); and
- investments in continuous improvement, to simultaneously reduce type I and type II errors.

The big question then is whether governments should progress toward the formulation of a single food protection agency, i.e. an agency that will coordinate information inflow and outflow or allocate scarce resources. Even within the United States, local, state, and federal inspection agencies face significant challenges with information sharing and in determining what information is acceptable. This might not be necessary given the complexity and scale of local, state, and federal inspection agencies. Intelligent systems (with advancements in technology) should be developed to enable agencies to develop a common database where information could be shared for inspection, research, and improvement of the system. To advance the science of real-time intelligent systems, research should be encouraged for the collection and dissemination of relevant data from a central location. This approach might provide a cost-effective solution to improve food safety, by reducing type I and type II food risks.

Lapses in policies and type I and type II food risks

Type I and type II food risks were introduced in Chapter 1. The interface model focuses on type II food risks, i.e. when a defective product is distributed to the consumer and causes harm that is extensive enough to create a market failure as a result of the failure to detect the problem or diagnose the cause. Type II disruptions are associated with the cost of illness or death from food safety problems. Examples of type II disruptions are failures in inspection systems for testing foodborne pathogens, food counterfeiting, and food adulteration. An obvious example of type II disruption was the *Salmonella* Saintpaul incidence (see Box 1.3 in Chapter 1). Another example of intentional adulteration was the melamine adulteration event in animal feed and milk powder (Xiu and Klein, 2010). Both of these examples led to multiple fatalities. The melamine adulteration episode resulted from an inspection system's failure to detect an intentional, although commercially motivated, set of actions by a large number of individuals. Nganje and Skilton (2011) noted that when information is shared effectively and efficient policies are designed, type I and type II food risks are minimized.

To assist policy makers in designing a system to detect, prevent, and respond to food safety/defense risks in the food supply networks, it is critical to understand a control-oriented system. While the emphasis of a control-oriented security system is to simultaneously minimize type I and type II errors, improvements can take the form of a reduction in one or both types of error with a subsequent reduction in the cost of microbial foodborne illnesses and deaths.

The main goal of the interface model or control-oriented system is to minimize controllable food risks in a continuous manner. This model differs significantly from the system designed to protect against disruptions caused by uncontrollable rare events such as food terrorism. One way that this difference can be understood is to note that error-based disruptions only occur if there is a detection and diagnostic process, as with the risk analysis model, intended to control a potentially disruptive defect or event (Meszaros, 1999; Lee and Wolfe, 2003). When detection or diagnostic systems fail, the normal function of the supply network delivers the defective product to the consumer. The problems from uncontrollable sources are distinct and further complicated by the complexity of the food supply network system (Nganje and Skilton, 2011).

Using the Interface Model to Identify Vulnerability in Orange Juice Processing

Sweet oranges were established in Florida between 1513 and 1565 with seeds imported from Europe (Food Source Information, 2015). Commercial oranges in Florida have continually been pushed further south to avoid fruit killing and freezes. From 1998 through 2012, orange juice has been implicated in ten reported outbreaks in the United States, during which 810 persons became ill with confirmed pathogenic agents like *Salmonella* and a Norwalk-like virus. Below, we identify the process oranges go through to be made into orange juice and the vulnerabilities associated with each process in this orange juice processing chain.

Farm operations

There must be adequate depth for good root development, and shallow soils with high water-holding ability should be avoided (Morton, 1987). Rootstocks affect the aldehyde content, and the oil content of the peel is influenced by the orange selection. Rootstock influences not only the rate of growth, disease resistance, and productivity of the cultivar, but also the physical and chemical attributes of the crop (Morton, 1987). Oranges are subject to a great number of diseases affecting the roots, trunk and branches, foliage, and fruit. The United States Environmental Protection Agency sets a maximum residue level for pesticides on fresh fruit. There is the possibility of excess chemical residue from fertilizers during production. Oranges should be monitored and treated before harvest to minimize defective fruits.

Harvesting operations

The 10-month harvest season starts in early October, and involves picking, loading, and packing the citrus fruits for transport to processing plants (Florida Agriculture, 2016). Oranges are harvested from large groves. When the mature fruit is ready to pick, a crew of pickers are sent in to pull the fruit off the trees (Advameg, 2016). Harvesters not following proper sanitation procedures can introduce pathogens through cross-contamination of oranges. Oranges can come into contact with feces or bird droppings, which can lead to contamination problems further down the processing chain.

Processing/grading operations

Grades for orange juice are specified by the United States Department of Agriculture (USDA). United States grades A and B are the quality of orange juice that meet the applicable USDA requirements. Substandard is the quality of orange juice that fails to meet the requirements for grade B (USDA AMS, 2016). A bad fruit that is not removed from the process could cause cross-contamination and increase the vulnerability of the end product, which could cause harm to humans.

Extraction of juice

Proper juice extraction is important to optimize the efficiency of the juice production process as well as the quality of the finished drink (Advameg, 2016). The extracted juice is filtered through a stainless steel screen before it is ready for the next stage. Cleaning and sanitation of the machinery needs to be implemented to prevent microbial growth, such as mold fungi and bacteria, which aids in the reduction of orange juice contamination.

Concentration of juice

The juice that is extracted is now mixed with water. Concentrated juice extract is approximately five times more concentrated than squeezed juice, and when diluted with water it is used to make frozen juice and many ready-to-drink beverages (Advameg, 2016). The water needs to be tested to ensure it is fit for drinking and free of microbial containments.

Reconstitution of juice

The next stop is the blending tanks that measure the natural sugar (Brix measurement) and, if necessary, blend in juice from other oranges to ensure a set level of sweetness (Florida Agriculture, 2016). Processors need be careful where they buy their additives so that their production and sources meet United States Food and Drug Administration (FDA) standards.

Pasteurization and elimination of potential food risks

Thanks to its low pH (about 4), orange juice has some natural protection from bacteria, yeast, and mold growth, but pasteurization is still required to further retard spoilage (Advameg, 2016). The incorporation of pasteurization into the production process has made fresh and frozen orange juice from concentrate safer to consume. However, the risk of drinking pasteurized orange juice remains as contamination by food handlers can still occur if it is not heated to temperature and time specifications.

Packaging/labeling

Packaging must be performed in a sanitary environment. Juice is packaged in cartons, bottles, and other containers stamped with an expiration date. To ensure sanitary conditions hot pasteurized juice should be filled in containers. Where possible, metal or glass bottles should be pre-heated (Advameg, 2016). Hydrogen peroxide or other approved sterilizing agents may be used if heat use is not feasible. This ensures that no contamination of any sort occurs as the orange juice goes to consumers.

Consumer

Consumers must store the orange juice at the right temperature to ensure microbial growth is minimal to none. Making sure this step is implemented will guarantee a nice glass of orange juice that is enjoyed without any resulting sickness.

Quality control

The quality is checked throughout the production process to ensure the orange juice is safe and wholesome for consumption. Inspectors grade the fruit before the juice is extracted and after extraction and concentration. The product is also checked to ensure it meets a number of USDA quality control standards. The final juice product is evaluated for a number of key parameters that include acidity, citrus oil level, pulp level, pulp cell integrity, color, viscosity, microbiological contamination, mouth feel, and taste (Advameg, 2016). Finally, the filling process units are inspected to make sure they are filled, sealed appropriately, and ready to be consumed. A final test for contaminents and pathogens needs to be implemented to minimize vulnerability.

Final control considerations and traceability

If there is an outbreak, it needs to be contained in a timely and targeted manner. Traceability is important so that a potential source of vulnerability can be identified and responded to accordingly in order to minimize food recalls and outbreaks (Stinson *et al.*, 2010). Identifying the problem quickly can help with a quicker recall, less consumption of defective orange juice, and fewer illnesses or death. If these vulnerabilities in the supply chain occur, the electronic barcode or radiofrequency identification could help pinpoint the vulnerability source (Stinson *et al.*, 2010).

Policy Response to Minimize Error-Based Disruptions

Our discussion from the interface model suggests that a lapse in policy increases the occurrence of microbial food risk to humans. The big question is why? Why are food safety risks not contained entirely by private firms? Microbial risks possess unique features that are different from other risks. For example, humans cannot see pathogens without the use of microscopes. This is called the credence attribute. Food safety is a public concern, and consumers demand a minimum level of safety be instituted in the system through control and testing. Food safety policies are present to minimize the voids in the private sector control measures. Box 4.1 provides an example of policy implementation to address an identified risk.

In Chapter 10, we discuss the detailed formulation of policy in the United States using the case of PR/HACCP. As causes can be complex (i.e. no point source) and contributing events can be dispersed across the food supply network, detection and prevention of unintentional error-based disruptions can be very difficult. Nganje and Skilton (2011) noted that a control-oriented process provides inspection, detection, trace, and prevention in the food supply, including error-based disruption points and subsequent opportunities for improvement of the risk analysis framework. It enables us to understand potential failure points in terms of risk, prevention, and safety, and then provides a framework to examine a response that can improve prevention and thus reduce risk while increasing food safety.

Box 4.2 provides examples of emerging and "difficult-to-control" food risks. We also present policy responses associated with these risks.

Synergies from multiple stakeholders

Nganje (2013) noted that challenges from globalization necessitate several agencies and stakeholders to address complex supply chain issues. For example, imported foods require the intervention of several agencies to control for food safety problems. These include the following:

- The USDA initiates inspections to comply with quality and grading standards at the farms in foreign countries and at their packing and processing facilities.
- Custom Border Protection (CBP) and the USDA work together to implement the 24-hour e-manifest rule

(US Department of Homeland Security, 2003). This rule requires logistics and transport entities to provide detailed descriptions of the contents of their shipment 24 hours before the commodity is transported. CBP also uses alternative forms of X-ray machines and gamma-imaging systems (risk-based technologies) to screen information on 100% of the cargo before it is shipped to the United States (Agriculture Protection Program, 2008). A shipment that meets the admissible protocol of the National Agriculture Release Program (NARP) is classified as a low-risk commodity.

- The FDA conducts pathogen testing at the port of entry. The FDA determines violations of incoming cargoes if pathogen performance standards are not met (US Food and Drug Administration, 2009).
- The Animal and Plant Health Inspection Service (APHIS) focuses on pest control. APHIS and Plant Protection and Quarantine regulations provide a list of all approved fruits and vegetables for imported foods.

Box 4.1. Voids in control measures in the United States

In 1996, the Food Safety Inspection Service (FSIS) introduced new mandatory food safety regulations following repeated discoveries of *Escherichia coli* and *Salmonella* in the United States food supply in the 1980s and early 1990s. The new regulations were called Pathogen Reduction/Hazard Analysis and Critical Control Points (PR/HACCP). They mandated the establishment of critical control points in food production and processing operations, and established testing routines for food products, while ensuring the safety of meat and poultry products. By the year 2000, these regulations had been adopted by meat and poultry processors. Pathogen levels have decreased since the adoption of mandatory PR/HACCP in meat and poultry processing (CDC, 2016). The CDC report shows a 30%, 9%, 32%, and 29% reduction in *Campylobacter*, *Salmonella*, *Listeria*, and *E. coli* O157, respectively, over the years. However, PR/HACCP was only mandated for meat and poultry processing, and is optional for other segments along the food supply chain.

Box 4.2. Policy response to mitigate emerging food risks

Investments in United States agroterrorism surveillance, preparedness, and response have heightened following the 11 September 2001 terrorist attack. Agricultural commodities travel through many forms of machinery, buildings, and storage facilities and along a variety of transportation and processing modes (from farm to wholesale, retail, and consumption). Any one of these places might be vulnerable to an agroterrorism attack. Risks associated with bioterrorism are an emerging food risk with important implications for many industries along the food supply chain. In December 2004, the FDA issued a Final Food Bioterrorism Regulation for the establishment and maintenance of records to track commodity flows one-step forward and one-step backward (OSF/OSB) for all firms along the food supply chain, including international food suppliers. However, there is no centralized system for information sharing that could increase communication and reduce lapses in the risk analysis model.

Exploring synergies helps reduce failure costs from type I errors or false positives. This also improves traceability. We expect food supply chain security managers to make two types of investments, depending on the nature of the diagnosis achieved through the interface model. First, consider continuous improvement efforts that can render the systems safer and more secure. Second, increase investments in threat-recognition systems. This kind of preventative investment helps to gather information about risk (Verduzco *et al.*, 2001). Threat-recognition systems make detection more accurate and therefore reduce vulnerabilities by reducing type II errors without increasing type I errors. A control framework for managers will also provide directions in finding ways to simultaneously mitigate threat and control the cost of errors. Box 4.3 describes a case of a false-positive result.

Limitations of Public–Private Partnership

Several international and national policies have been developed and implemented to improve food safety and quality over the years. Some notable provisions include the Sanitary and Phytosanitary measures by the World Trade Organization, the North American Free Trade Agreement, (NAFTA) and most recently, the Food Safety Modernization Act (FSMA). One major component of this provision is compliance verification. The FSMA provides the USDA's FSIS more authority on ensuring food safety from domestic and imported sources. The law requires that importers perform risk-based verification analyses to assure that foods are produced in compliance with HACCP procedures (see Chapter 10) and are not adulterated or misbranded, and that foreign facilities will operate in a manner that ensures compliance with United States food safety standards. The law also requires that a traceability system be adopted. However, coordination and information sharing continue to pose significant challenges.

Summary

The risk analysis model provides a framework for stakeholders to channel information to policy makers, to design efficient control, and to minimize food risks. The interface model could be used to explain how microbes become risks to humans with type II disruptions that have consequences such as making people sick or killing them. Private and public partnerships and information sharing are important ways to mitigate foodborne illness risks and costs.

References

Advameg (2016) Orange juice. Available at: http://www.madehow.com/Volume-4/Orange-Juice.html (accessed 28 August 2016).

Agriculture Protection Program (2008) Frontline. Available at: https://www.cbp.gov/sites/default/files/documents/frontline_vol1_issue2.pdf. (accessed 10 October 2016).

Behravesh, C.B., Mody, R.K., Jungk, J., Gaul, L., Redd, J.T., Chen, S., Cosgrove, S., Hedican, E., Sweat, D., Chavez-Hauser, L., Snow, S.L., Hanson, H., Nguyen, T.A., Sodha, S.V., Boore, A.L., Russo, E., Mikoleit, M., Theobald, L., Gerner-Smidt, P., Hoekstra, R.M., Angulo, F.J., Swerdlow, D.L., Tauxe, R.V., Griffin, P.M., and Williams, I.T. (2011) 2008 outbreak of *Salmonella* Saintpaul infection associated with raw produce. *New England Journal of Medicine* 364, 918–927.

Bohn, R.E. (1994) Measuring and managing technological knowledge. *Sloan Management Review* 36, 61–73.

CDC (2016) Foodborne Diseases Active Surveillance Network (FoodNet), United States Centers for Disease Control. Available at: http://www.cdc.gov/foodnet/index.html. (accessed 1 March 2016).

Florida Agriculture in the Classroom, Inc. (2016). From Grove to You. Available at http://faitc.org/wp-content/uploads/2013/07/From-Grove-to-You-Information1.pdf (accessed 13 January 2017).

Flynn, B. (2013) Tomato growers want compensation for losses in 2008 outbreak. *Food Safety News*. Available at http://www.foodsafetynews.com/2013/08/tomato-growers-want-to-be-compensated-for-losses-in-2008-outbreak/#.V4AHO_krLmE (accessed 30 December 2015)

Food Source Information (2015) A food production wiki for public health professionals. Available at http://fsi.colostate.edu/oranges/ (accessed 28 August 2016).

Jensen, H.H. and Unnevehr, L. (2009) HACCP in Pork Processing: Cost and Benefits, Working Paper 99-WP 227.

Lee, H.L. and Whang, S. (2005) Higher supply chain security with lower cost: Lessons from total quality management. *International Journal of Production Economics* 96, 289–300.

Lee, H.L. and Wolfe, M.L. (2003) Supply chain security without tears. *Supply Chain Management Review* 7, 12–20.

Meszaros, J.R. (1999) Preventive choices: organizations' heuristics, decision processes and catastrophic risks. *Journal of Management Studies* 36, 977–998.

Morton, J. F. (1987) Orange. Available at: https://www.hort.purdue.edu/newcrop/morton/orange.html (accessed 28 August 2016).

Nganje, W. (2013) Food import safety and quality: food safety and food defense risks. In: *Improving Food Import Safety and Trade.* USDA-Markets and Trade.

Nganje, W.E. and Skilton, P. (2011) Food risks and type I and II errors. *International Food and Agribusiness Management Review* 14, 109–123.

Stinson, T., Mejia, C., McEntire, J., Keener, K., Muth, M.K., Nganje, W., and Jensen, H. (2010) Traceability (product tracing) in food systems: an IFT report submitted to FDA, Volume 2: Cost considerations and implications, *Comprehensive Reviews in Food Science and Food Safety* 9, 159–175.

USDA AMS (2016) Orange juice from concentrate grades and standards. Available at https://www.ams.usda.gov/grades-standards/orange-juice-concentrate-grades-and-standards (accessed 28 August 2016).

US Department of Homeland Security (2003) Federal Register, Part II. Available at: http://www.census.gov/foreign-trade/regulations/fedregnotices/ADVANCEFILING.pdf. (accessed 10 October 2016).

US Food and Drug Administration (FDA) (2009) Safety. Available at: http://www.fda.gov/Safety/Recalls/ucm165546.htm. (accessed 10 October 2016).

Verduzco, A., Villalobos, J.R., and Vega, B. (2001). Information-based inspection allocation for real-time inspection systems. *Journal of Manufacturing Systems* 20, 13–22.

WHO (2015) About risk analysis in food. Available at: http://www.who.int/foodsafety/risk-analysis/en/ (accessed 26 February 2015).

WHO (2016) Food safety: risk assessment. World Health Organization. Available at: http://www.who.int/foodsafety/risk-analysis/riskassessment/en/ (accessed 1 March 2016).

Xiu, C. and Klein, K.K. (2010) Melamine in milk products in China: examining the factors that led to deliberate use of the contaminant. *Food Policy* 35, 463–470.

5 Foodborne Infections, Intoxications, and Etiology

> **Key Questions**
> - What is a foodborne infection?
> - What is a foodborne intoxication?
> - What is the etiology of foodborne disease?
> - What are the basics of testing foods for foodborne microorganisms?

Foodborne Illness

Food poisoning is a colloquial term for illness due to consuming food. Foodborne illness is a more appropriate term and includes any human injury or disease state induced as a consequence of consuming a particular food or beverage. The causes are the foodborne hazards, which can be chemical (i.e. toxins), physical (i.e. choking hazards, sharp objects, stones), or biological in nature. "Unsafe food containing harmful bacteria, viruses, parasites or chemical substances causes more than 200 diseases – ranging from diarrhoea to cancers" (WHO, 2015). We will focus our discussion on the biological hazards presented by foodborne microorganisms.

In Chapter 4, the concept of risk was introduced. Risk levels due to biological hazards in foods can be managed to keep risk levels low. However, risk can never be controlled so completely that there is zero risk. Safety is an absolute concept, and in reality there are no absolutes. A food cannot be proven safe (Peterson, 2005). Risk levels can never be so controlled that there is no risk, but it is still important to find ways to lower risk, especially for foods at high risk of transmitting illness.

Consumers find the idea of microorganisms being in their food to be disconcerting. The vast majority of foods are not truly sterile, even after cooking and processing. We don't need to fear all microorganisms in our food. Some are innocuous or even beneficial to health, i.e. probiotics. Some can multiply and cause food to spoil if conditions support their growth. Food spoilage is a defect of quality, and not a safety concern. The microorganisms of concern for food safety are the pathogens, not the spoilage microorganisms. To summarize, foods can contain beneficial microorganisms, spoilage microorganisms, and, more rarely, pathogens – the good, the bad, and the ugly (see Box 5.1). For an interesting take on this concept with cheese, where these delineations are not always clear, see Marcellino and Benson (2013).

What is a pathogen? Casadevall and Pirofski (2002) define a pathogen as "a microbe capable of causing host damage." The capability of a microbe to cause host damage is described in terms of virulence. Virulence levels depend on both the microorganism and the host. Host susceptibility plays an important role, and hosts with impaired immune systems are at higher risk of harm from microorganisms.

Immunocompromised is the term used for hosts with impaired immune systems. Examples of people in immunocompromised states include:

- the very old, for whom immune systems are failing to function as well;
- the very young, for whom immune systems are not yet fully developed;
- those with chronic diseases such as cancer or diabetes;
- people with organ transplants who take medications to prevent tissue rejection;

- pregnant women whose immune systems are taxed due to the biological burden of pregnancy; and
- nutrient-deprived people.

Consider how many people you know who fit with these examples. This status describes a large percentage of the human population.

Foodborne Infections

Foodborne infections are a form of disease that results from eating food containing living pathogens.

An infection involves the invasion by living microorganisms of tissue in the host. This invasion and multiplication process causes damage to the host. An active infection can be localized or can spread through blood or lymph and become systemic, affecting multiple locations and tissues in the host. The incubation period is the time period between consumption (exposure) and development of symptoms.

Microorganisms in a non-harmful state within the host are not considered to be an infection. The microbiome of the gut would not be considered an infection. However, opportunistic microorganisms can cause infections when the correct conditions are presented.

Foodborne infections involve the transmission of living pathogens from a food or beverage to a host. In order for the pathogen to successfully invade the host's tissue, it must first survive the processes of host ingestion and digestion, and escape elimination. Ingestion involves the physical mastication of food and pre-digestion with enzymes found in saliva. After food is swallowed, it enters the stomach where conditions of very low pH and further enzymatic digestion occur. After leaving the stomach, a pathogen would need to survive a dramatic change in pH, further enzymatic action, and competition with other microorganisms for places to attach in the intestinal tract, or be expelled with fecal material. Even if a microbe survives all of the barriers of the gastrointestinal tract and successfully attaches to host tissue, it must then still survive the host's immune system (Tomasello and Bedoui, 2013). The complex interaction of the gut microbial biome, epithelial cells, and the host immune system present multiple hurdles that a pathogen needs to overcome to be a successful pathogen.

Virulence is related to a pathogen's ability to overcome host defenses and cause host damage. An infective dose is the number of specific microorganisms needed to cause an infection, and virulence affects that threshold number. Typically, the more virulent the pathogen, the lower the infective dose.

A host can carry potentially pathogenic microorganisms, but not have an active infection. This is called an asymptomatic carrier state and can be an important point of control for infection prevention. See Box 5.2 and Fig. 5.1 for details of an infamous asymptomatic carrier. In Chapter 6 to Chapter 9, we will further discuss how specific pathogens vary in virulence and infectious dose, and how they may be carried and transmitted.

Foodborne Intoxications

A foodborne intoxication is a state where there is damage to a consumer (host) from a consumed chemical. Toxicity is the intrinsic capacity of a chemical to produce injury. A toxin is a substance that has toxicity. Toxicity is limited by dose, and even water is toxic at a high enough dose. Most potent toxins require very low doses to cause damage.

Box 5.2. Typhoid Mary

There once was a cook named Mary Mallon. She was born in Ireland in 1869 and emigrated to the United States. She was also the first person ever in the United States to be identified as an asymptomatic carrier of infectious disease (Brooks, 1996).

In 1906, Mallon was working as a cook in a house in Oyster Bay, Long Island. The home had been rented by a banker's family. There were four members of the family and seven servants in the house. About 3 weeks after Mallon started working as a cook there, six of these inhabitants came down with typhoid fever, a very serious disease with a mortality rate of about 10% at that time. Typhoid fever is caused by *Salmonella* Typhi, with an incubation period of about 3 weeks.

George Soper, a New York Department of Health investigator, was hired by the owner of the house to try to determine the cause. He suspected the source to be the cook due to the incubation period and Mallon's records of employment. Most of the homes at which she had worked as a cook had been stricken with typhoid fever.

> It was to be Mary Mallon's fate to clear away much of the mystery which surrounded the transmission of typhoid fever and to call attention to the fact that it was often persons rather than things who offered the proper explanation when the disease occurred in endemic, sporadic and epidemic form (Soper, 1939).

When Soper confronted Mallon, offering free medical care and a request that she change her profession, she chased him out of the house armed with a carving knife. Mallon was not convinced that she carried an infectious disease since she showed no symptoms, and had no desire to change her profession. The Board of Health, on Soper's informing, forcibly detained Mallon after some intense struggle. They loaded her into an ambulance and brought her to a hospital, where it was confirmed that she was shedding *Salmonella* Typhi. By that time, she had infected 22 people, including a little girl who had died from typhoid fever.

Soper told Mallon she could be released if she would stop working as a cook and have her gall bladder removed (believed to be the source of the infection at the time). "She denied that she was responsible for anyone's sickness or death and refused to recognize the authority of science or government to label her a menace to society" (Brooks, 1996). She was subsequently quarantined in a one-room bungalow at the Riverside Hospitable for Communicable Diseases.

After 3 years of confinement, a new health commissioner agreed to release Mallon if she promised not to work as a cook. She changed her identity and worked as a cook over the following 5 years. She was caught working as a cook in a maternity hospital that had 25 new cases of typhoid.

In 1907, she was subdued by force and sent back to her place of quarantine. Mary Mallon died on 11 November 1938 after 26 years of forced isolation. She likely infected at least 51 people, three of whom died from typhoid fever. "As a healthy carrier of *Salmonella typhi* her nickname of 'Typhoid Mary' had become synonymous with the spread of disease" (Marineli *et al.*, 2013).

Her case remains one of study for both infectious disease control and the ethical treatment of carriers. "Mary Mallon's intransigence had temporarily made society's right to protect itself more important than the liberty of one person" (Brooks, 1996).

Fig. 5.1. An illustration of Typhoid Mary that appeared in the 20 June 1909 issue of *The New York American* (public domain).

In terms of microbial food safety, these toxic chemicals are typically produced by foodborne microorganisms. Microbial toxins tend to be secondary metabolites (see Chapter 2). Toxicity is dependent on dosage, and in order for a foodborne intoxication to occur, the microorganisms must have produced a sufficient amount of toxin in the food prior to consumption. This type of foodborne illness is most common in foods that have not been stored at the proper temperature, or in other words, temperature abuse has occurred.

Botulism is a type of foodborne intoxication that many people have heard of. This is caused by a very potent toxin produced by strains of *Clostridium botulinum*. The botulism toxin is a powerful neurotoxin, and intoxication with this toxin can be lethal. When conditions are right for the growth of this bacteria in a food system, it does not take much growth to produce a lethal amount of neurotoxin.

Most microbial foodborne toxins are produced by specific types of bacteria or fungi. The most common types of bacterial foodborne toxins are produced by:

- *Clostridium botulinum*;
- *Staphylococcus aureus*; and
- *Bacillus cereus*.

Many fungi produce toxic substances, and these include toxic yeast by-products, poisonous mushrooms, and mold toxins referred to as mycotoxins. Mycotoxins lead to some of the largest food losses on a global scale by making food unsafe for consumption. The mold genera of most concern for mycotoxin production in agricultural commodities include:

- *Aspergillus*;
- *Penicillium*; and
- *Fusarium*.

Several other genera are also of concern. See Chapter 6 and Chapter 8 to learn more about the specific microorganisms of highest concern for producing foodborne toxins.

Toxin-Mediated Infections

Toxin-mediated infections, also referred to as toxico-infections, involve ingestion of large numbers of live pathogens, which then produce or release toxin in the gastrointestinal tract. These toxins tend to cause host damage that gives the pathogen an advantage for forming an infection. Large numbers are required for the pathogen dose as enough must be able to bind to or invade the host to cause enough damage to advantage establishment of an infection. Foodborne pathogens of most concern for toxin-mediated infections include:

- *Bacillus cereus*;
- *Clostridium perfringens*;
- enterotoxigenic and enterohemorrhagic *Escherichia coli*;
- some strains of *Campylobacter jejuni*; and
- *Vibrio cholera*.

These tend to be problems in temperature-abused foods or foods highly contaminated with fecal material. The various mechanisms that these pathogens use to damage the host will be discussed in the following chapters.

Etiology and Epidemiology of Foodborne Illnesses

Etiology is defined by the Oxford English Dictionary as "The cause, set of causes, or manner of causation of a disease or condition." The tool often used to determine etiology is epidemiology, defined as "The branch of medicine that deals with the incidence, distribution, and possible control of diseases and other factors relating to health." Determining the cause of foodborne illness is often not an easy task (see Box 5.2 for an historical perspective). Gould *et al.* (2013) estimate 48 million illnesses each year in the United States. Most of these go unreported. Many of the foodborne illnesses reported in the United States have no verified cause.

In the United States, about 38% of documented outbreaks are of unknown etiology (Gould *et al.*, 2013). Of the reported illnesses, about 9.4 million were caused by 31 known pathogens (Scallon *et al.*, 2011). Of those that were known, most cases of foodborne illnesses were caused by norovirus (58%), non-typhoidal *Salmonella* spp. (11%), *Clostridium perfringens* (10%), and *Campylobacter* spp. (9%). The most common causes of hospitalization were non-typhoidal *Salmonella* spp. (35%), norovirus (26%), *Campylobacter* spp. (15%), and *Toxoplasma gondii* (8%). The most common causes of death were non-typhoidal *Salmonella* spp. (28%), *T. gondii* (24%), *Listeria monocytogenes* (19%), and norovirus (11%).

The United States Centers for Disease Control and Prevention has established a network of public health laboratories to subtype microorganisms linked to foodborne illness outbreaks. This network is called PulseNet (CDC, 2015). The labs that are in the PulseNet use standardized testing methods that allow the DNA fingerprint of the microbe to be taken and compared to other cases occurring around the country. This approach has been an effective public health intervention to more quickly contain outbreaks that cross state lines. PulseNet International does similar partnering across 82 countries.

The information about the etiologies of foodborne illnesses are an important aspect of epidemiology. This information helps to inform policy makers and educators on how to implement preventative measures to lower the overall risk levels of foodborne disease. It helps guide research initiatives for understanding these pathogens and developing innovations to enhance prevention as well as treatment for those affected.

Testing Methods for Pathogens in Foods

Microbiological methods for detecting and measuring foodborne pathogens and toxins are vital tools for determining etiology and control methods for foodborne illness. Microbiological methods range from traditional culturing techniques to more modern molecular-based and rapid methods. Methodology selection depends on needs for sensitivity, selectivity, accuracy, precision, rapidity, ruggedness, usability, and cost. "Ideal methods would identity and quantify multiple targets from any matrix in real time under non-laboratory conditions, and cost little" (Ryu and Wolf-Hall, 2014). Most testing labs are shifting from traditional culturing methods to rapid methods to save on labor costs and turnaround time.

No method is perfect, and it is important to understand the types of errors that limit how results are used. Limitations include:

- type I errors related to a true negative testing positive; and
- type II errors related to a true positive testing negative.

The AOAC International (AOAC used to stand for Association of Official Agricultural Chemists; the organization now simply uses AOAC International) is an organization focused on uniformity of testing methods. For a detailed discussion of setting testing standards, see the AOAC International Presidential Task Force on Best Practices in Microbiological Methodology (AOAC International, 2006). European organizations that validate standard methods include the European Certification Organization for the Validation and Approval of Alternative Methods for the Microbiological Analysis of Food and Beverages (MicroVal) and the Association Français de Normalisation (Valderrama et al., 2015).

Standard methods manuals are referred to for ensuring consistency of methods. The United States Food and Drug Administration's Bacteriological Analytical Manual (BAM) is a freely available manual and is followed by regulatory bodies. "The FDA sets scientific standards for testing foods for various contaminants. Laboratories and food companies worldwide use these standards to make sure that food products are safe to eat and drink" (FDA, 2015). The BAM website lists other microbiological methods information and resources including:

- American Public Health Association, which includes the Compendium of Methods for the Microbiological Examination of Foods (Doores et al., 2013);
- American Type Culture Collection, which is a culture collection including many reference strains used in testing methods (www.atcc.org/);
- AOAC International Official Methods of Analysis (www.eoma.aoac.org/);
- Centers for Disease Control and Prevention - Diagnosis and Management of Foodborne Illnesses: A Primer for Physicians (www.cdc.gov/mmwr/preview/mmwrhtml/rr5002a1.htm);
- Health Protection Branch, Health and Welfare Canada – Compendium of Analytical Methods (www.hc-sc.gc.ca/fn-an/res-rech/analy-meth/microbio/index-eng.php);
- International Organization for Standardization (ISO) (www.iso.org/iso/.htm); and
- Food Safety and Inspection Service, United States Department of Agriculture, USDA/FSIS Microbiology Laboratory Guidebook (www.fsis.usda.gov/wps/portal/fsis/topics/science/laboratories-and-procedures/guidebooks-and-methods/microbiology-laboratory-guidebook/microbiology-laboratory-guidebook).

Rapid methods are commonly used, and many rapid methods kits are available commercially. New innovations appear on the market with increased frequency.

Valderrama *et al.* (2015) provide a review of commercially available rapid methods for select foodborne pathogens including *Listeria monocytogenes*, *Salmonella* spp., *Staphylococcus aureus*, and Shiga toxin-producing *Escherichia coli*. This review describes immunology-based, nucleic acid-based, and biosensor methods. Culture enrichment steps are still common and indicate the limitations in sensitivity that these methods can achieve. Other limitations include differentiation between viable and non-viable cells and detection of viable but non-culturable cells. Sampling is described as a critical step:

> The utilization of commercial rapid detection methods demands that established sampling procedures are strictly followed. This approach is crucial because interpretations derived from a large batch of food may be based on a relatively small sample. If a sample is improperly collected, mishandled, contaminated, or is not representative, the results can be meaningless or worse, misleading (Valderrama *et al.*, 2015).

In the following chapters describing specific microbiological hazards, common testing methods for specific target microorganisms will be referenced.

Summary

Foodborne infections result from consumption of living virulent pathogens. Foodborne intoxications result from consumption of toxins produced by microorganisms in food prior to consumption. Toxin-mediated infections are infections caused by pathogens that excrete toxins to facilitate establishment of infection. All of these are dose dependent, requiring a threshold number of pathogens be consumed or a threshold amount of toxin be eaten before the host experiences damage.

The etiology of foodborne disease is the cause of the disease. Epidemiology is a tool that examines disease outbreaks in order to determine the etiology. Testing methods are a critical part of this methodology to link the foods consumed to the victims.

Further Reading

Banerjee P. and Fung, D.Y.C. (2016) *Rapid Methods in Food Microbiology*. CRC Press, Oxford, UK.

Doyle, M.P. and Beuchat, L.R. (2007) *Food Microbiology: Fundamentals and Frontiers*, 3rd edn. ASM Press, Washington, DC, USA.

Jarrow, G. (2015) *Fatal Fever: Tracking Down Typhoid Mary*. Calkins Creek, Honesdale, PA, USA.

Juneja, V.K. and Sofos, J.N. (2010) *Pathogens and Toxins in Foods: Challenges and Interventions*. ASM Press, Washington, DC, USA.

References

AOAC International (2006) Final report and executive summaries from the AOAC International presidential task force on best practices in microbiological methodology. Available at: http://www.fda.gov/Food/FoodScienceResearch/LaboratoryMethods/ucm124900.htm (accessed 15 March 2015).

Brooks, J. (1996) The sad and tragic life of Typhoid Mary. *Canadian Medical Association Journal* 154, 915–916.

Casadevall, A. and Pirofski, L. (2002) What is a Pathogen? *Annals of Medicine* 34, 2–4.

CDC (2013) Tracking and reporting foodborne disease outbreaks. Available at: http://www.cdc.gov/Features/dsFoodborneOutbreaks/ (accessed 5 April 2015).

CDC (2015) PulseNet. Available at: http://www.cdc.gov/pulsenet/ (accessed 5 April 2015).

Doores, S., Slafinger, Y., and Tortorello, M. (2013) *Compendium of Methods for the Microbiological Examination of Foods*, 5th edn. American Public Health Association, Washington, DC, USA.

FDA (2015) Microbiological Methods & Bacteriological Analytical Manual (BAM). Available at: http://www.fda.gov/Food/FoodScienceResearch/LaboratoryMethods/ucm114664.htm (accessed 15 March 2015).

Gould, L.H., Walsh, K.A., Vieira, A.R., Herman, K., Williams, I.T., and Cole, D. (2013) Surveillance for foodborne disease outbreaks – United States, 1998–2008. *Morbidity and Mortality Weekly Report* 62, 1–34.

Helms, M., Vastrup, P., Gerner-Smidt, P., and Molbak, K. (2003) Short and long term mortality associated with foodborne bacterial gastrointestinal infections: registry based study. *British Medical Journal* 326, 357–359.

Hoffmann, S, Batz, M.B., and Morris, J.G. (2012) Annual cost of illness and quality-adjusted life year losses in the United States due to 14 foodborne pathogens. *Journal of Food Protection* 75, 1229–1302.

Marcellino N. and Benson, D.R. (2013) The good, the bad, and the ugly: tales of mold-ripened cheese. *Microbiology Spectrum* 1, 1–27.

Marineli, F., Tsoucalas, G., Karmanou, M., and Androutsos, G. (2013) Mary Mallon (1869–1938) and the history of typhoid fever. *Annals of Gastroenterology* 26, 132–134.

Peterson, R.K.D. (2005) Why scientists can never prove that biotech crops are safe. Available at: http://agbiosafety.unl.edu/science.shtml (accessed 15 March 2015).

Ryu, D. and Wolf-Hall, C. (2014) Yeasts and molds. In: Doores, S., Slafinger, Y. and Tortorello, M. (eds) *Compendium of Methods for the Microbiological Examination of Foods*, 5th edn. American Public Health Association, Washington, DC, USA.

Scallon, E., Hoekstra, R.M., Angulo, F.J., Tauxe, R.V., Widdowson, M.A., Roy, S.L., Jones, J.L., and Griffin, P.M. (2011) Foodborne illness acquired in the United States – major pathogens. *Emerging Infectious Diseases* 17, 7–15.

Soper, G. (1939) The curious career of typhoid Mary. *Journal of Urban Health: Bulletin of the New York Academy of Medicine* 15, 698–712.

Tomasello, E. and Bedoui, S. (2013) Intestinal innate immune cells in gut homeostasis and immunosurveillance. *Immunology and Cell Biology* 91, 201–203.

Valderrama, W.B., Dudley, E.G., Doors, S., and Cutter, C.N. (2015) Commercially available rapid methods for detection of selected foodborne pathogens. *Critical Reviews in Food Science and Nutrition* 56, 1519–1531.

WHO (2015) *Food Safety*. World Health Organization, Fact sheet No. 399, March. World Health Organization, Geneva.

6 Gram-Positive Bacteria

Key Questions
- Which Gram-positive bacteria are of most concern for microbial food safety?
- What are the mechanisms by which these Gram-positive bacteria cause illness?
- What are the hazards these Gram-positive bacteria present to consumers?
- What controls are available to prevent foodborne illness due to these Gram-positive bacteria?

The Difference Between Gram-Positive and Gram-Negative Bacteria

Hans Christian Gram was the Danish scientist who, in 1884, published a method for a staining technique to help better see bacteria in tissue samples under the microscope. An unanticipated result of this technique was a way to differentiate two major groups of bacteria based on their cell wall compositions.

Bacterial cell membranes that contain thick layers of peptidoglycan are able to retain the crystal violet stain used in the method, resulting in purple- or violet-stained cells that are described as Gram positive. Bacteria that contain less peptidoglycan in their cell membranes are unable to retain the crystal violet stain after the destaining step of the procedure, and as a result of counterstaining with safranin dye appear red or pink, and are described as Gram negative. As with all microbiological testing methods, there are limitations and some bacterial species may produce Gram-variable results, indicating an ability to stain with either reaction result and not provide a clear distinctive result.

Gram staining is a preliminary test used on bacterial cultures to give clues to the identity of the species. The Gram reaction is a first clue, and then other clues like the microscopic cell morphology or shape and placement can provide other clues. The remainder of this chapter will focus on those bacterial pathogens of most concern in foods that fall under the category of Gram positive, and Chapter 7 will focus on those that are Gram negative.

Gram-Positive Bacteria of Concern in Food Safety

The following descriptions further classify Gram-positive bacteria of concern for food safety into two additional categories: spore formers and non-spore formers.

The spore formers

Spore-forming bacteria are all Gram-positive rods (see Fig. 6.1). Spore-forming bacteria have the capability to alter their cells to more hardy forms, or spores that can survive extreme environmental conditions. Spores are dormant and essentially biologically inactive. Bacterial spores are of most concern in foods where the food has been processed previously, reducing the competitive microbial flora. After processing, if the environmental conditions are favorable, the spores can germinate; the bacteria become biologically active and can multiply rapidly. For a deeper understanding of the biology of spore-forming bacteria, see Setlow and Johnson (2013). The bacterial spore formers of most concern for foodborne illness include *Bacillus cereus*, *Clostridium botulinum*, and *Clostridium perfringens*.

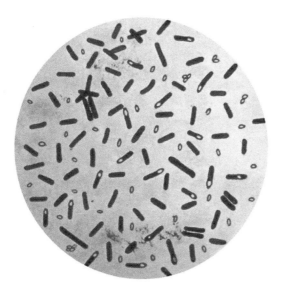

Fig. 6.1. Illustration of culture specimens of *Clostridium botulinum* depicting the rods of spore-forming bacteria that sometimes appear to have bulges where the spores are forming (CDC PHIL, 2014).

Bacillus cereus

Microscopic morphology: This species is a Gram-positive spore former, with vegetative cells that appear as large rods. The genus name *Bacillus* is derived from the Latin word bacillum, which means staff or walking stick. The species name *cereus* means wax or candle in Latin.

Environmental factors: As with most spore-forming bacteria, these species are ubiquitous in nature and can be found in almost any environment where there is soil and dust. Therefore, it is common to detect low levels of this species in many types of foods. Foodborne-illness risk increases as conditions become conducive to rapid multiplication of this species in a food system.

This species is a thermoduric mesophile with growth supported by temperatures ranging from 4 to 50°C (40–122°F), and an optimum growth temperature of 30°C (86°F). This means that this species can grow very slowly at either end of these temperature ranges. At an optimum temperature, this species can grow very quickly, with a generation time of 20 to 30 minutes. The spores of *B. cereus* are heat resistant and may survive cooking and then germinate and multiply post-heat processing, if environmental conditions such as temperature are favorable.

B. cereus is aerobic, and requires oxygen to multiply. This species can multiply in a range of pH conditions, from 4.9 to 9.3.

Type of foodborne illness: This species causes intoxications that are typically mild and self-limited. *B. cereus* can produce two different toxins. These toxins are described as a diarrheal toxin and an emetic toxin. In order for there to be a symptom-inducing dosage of either toxin, a concentration of vegetative cells needs to reach about 10^3 per gram of food.

The diarrheal toxin is released from live vegetative bacterial cells in the mode of a toxin-mediated infection. In other words, a large number of live bacteria must be ingested for a high enough dosage of toxin to be released from the cells to induce symptoms. The toxin is released from the bacterial cells as they are digested in the small intestine. The toxin acts upon the intestinal tract, inducing diarrhea. The diarrhea is typically watery, results in abdominal pain, and is fairly mild. The onset time from ingestion to symptoms is 8 to 16 hours, and the duration of the symptoms is typically 12 to 24 hours. The syndrome is similar to that caused by *C. perfringens*.

The emetic toxin is produced by the bacteria as it grows in the food. This toxin is heat stable and can remain stable after a food is heated. This toxin causes a true intoxication once a high enough dose is consumed. The onset time from consumption to symptoms forming is only 0.5 to 5 hours. The symptoms include nausea, vomiting, and malaise, and can also include diarrhea. The duration of symptoms is typically 6 to 24 hours. The emetic toxin causes a more severe and acute intoxication than does the diarrheal toxin. The emetic toxin syndrome is similar to that caused by *Staphylococcus aureus*. See Box 6.1 for a description of fried rice syndrome.

Foodborne illness statistics: *B. cereus* is estimated to cause 63,000 illnesses, 20 hospitalizations, and zero deaths per year in the United States (Scallon *et al.*, 2011).

Costs of foodborne illness: US$234/case in the United States in 2010 (Scharff, 2012). This was the lowest cost foodborne illness pathogen of the study. This is due to both a low mortality rate and the milder self-limited nature of the form of illness compared to other foodborne pathogens.

Outbreaks of foodborne illnesses linked to rice, especially fried rice, are often attributed to *B. cereus*, and in particular, the emetic toxin produced by this bacterial species. Cooked rice, and similarly starchy foods, that are temperature abused are an excellent medium for growth and toxin production by *B. cereus* (Agata *et al*., 2002).

When groups of people are suddenly hit with extreme nausea and vomiting, it is important to determine the cause as quickly as possible.

Diarrhea and emetic complaints from patients are encountered by emergency care providers on a daily basis. The goal of providers is more to categorize the patient's condition into one of the major types rather than to find a specific cause, all while attempting to identify the life threats associated with the condition and treating them appropriately (Asaeda *et al*., 2005).

This can be challenging when many things can cause these symptoms, ranging from terroristic intentional poisonings to unintentionally contaminated food. Asaeda *et al*. (2005) described a case study of a basic life support emergency medical technician unit being dispatched to assess a call from a 13-year-old girl reporting that her mother was violently ill. Upon arrival, both the mother and the grandmother were vomiting. The crew immediately investigated to determine if carbon monoxide poisoning might be the cause as this was a cold day and malfunctioning heating units can cause this. This was basically ruled out since the 13-year-old girl did not have symptoms. Upon interviewing the two with symptoms and subsequent follow-up, it was determined that the two victims had consumed leftover pork fried rice for lunch that day. The food had been left out overnight, unrefrigerated. This scenario and condition are so commonly associated with fried rice that the term "fried rice syndrome" has been used in the literature.

Testing methods: Culturing techniques for detecting and estimating numbers of *B. cereus* in a food sample are described by Tallent *et al*. (2012). Mannitol egg yolk polymixin (MYP) agar is one of the most common microbial media for enumeration of presumptive *B. cereus* (Bennett *et al*., 2015b). Concentrations of *B. cereus* need to be at least 10^5 per gram to implicate it as a causative agent, since a high dosage is needed to induce symptoms. Also described are the limitations of these methods. Bennett (2001) has described the methodology for determining enterotoxigenicity of isolated bacteria.

Controls: Effective controls to lower the risk of foodborne illness from this species include the following:

- Good sanitation practices will limit the number of spores present in a food.
- Temperature control that keeps food out of the temperature danger zone (4–60°C; 40–140°F) is important to prevent growth and toxin production. This does not allow time and temperature for the low background levels of this bacterial species to multiply. It is important to portion large volumes of foods in shallow containers to facilitate rapid cooling.

- Acidic foods with pH values lower than 4.6 are low risk for illness due to this pathogen.
- Preservatives, such as sorbates, in combination with controlling environmental factors can prevent the growth of *B. cereus*.

Clostridium species

There are two main species of concern for foodborne illness within the genus *Clostridium*. These include *C. botulinum* and *C. perfringens*. The genus name *Clostridium* comes from the Greek word klōstērm, which means spindle. The rod-shaped cells tend to have bulges where the endospores form, resembling the shape of a spindle.

CLOSTRIDIUM BOTULINUM

Microscopic morphology: This species is a Gram-positive spore former, with a spindle rod shape.

Environmental factors: As with most spore-forming bacteria, these species are ubiquitous in nature and can be found in almost any environment where there is soil and dust. Therefore, it is common to detect low levels of this species in many types of foods. Foodborne-illness risk increases as conditions become conducive to rapid multiplication of this species in a food system.

This species is a thermoduric mesophile with growth supported by temperatures ranging from 4 to 50°C (40–122°F), and an optimum growth temperature of 37°C (98.6°F, human body temperature). There are strains that have psychotrophic characteristics that increase foodborne-illness risk in refrigerated, hermetically sealed, low-acid foods such as vacuum-packaged seafood. The spores are heat resistant, may survive cooking, and then germinate and multiply post-heat processing if environmental conditions are favorable.

C. botulinum is a strict anaerobe, and oxygen is toxic to it. This species can multiply at pH values higher than 4.5 and requires many nutrients to grow. It does not grow well in the presence of competitive microflora. It also requires a high water activity (a_w) of at least 0.94 to grow.

Type of foodborne illness: This species causes a life-threatening intoxication. *C. botulinum* produces an extremely powerful neurotoxin that is formed in the food prior to consumption. The neurotoxin is a protein and is heat labile. Very little growth is needed to reach a dosage that can be toxic. The neurotoxin binds irreversibly to neural receptors, and results in the paralysis of muscles. When the diaphragm muscle is paralyzed, death due to asphyxiation occurs. Onset time for symptoms ranges from 18 to 36 hours. The mortality rate is high for botulism, and if survived, recovery can take months. Fortunately, outbreaks tend to be small in size. See Box 6.2 for examples of outbreaks with public health implications.

Very rarely, *C. botulinum* spores can cause a toxico-infection in infants who consume spores but do not yet have fully colonized digestive tracts. The spores can germinate within the gastrointestinal tract and excrete the neurotoxin. This condition is called infant botulism. Honey has been linked to infant botulism, although other sources of spores may be responsible as well.

Foodborne illness statistics: *C. botulinum* is estimated to cause 55 illnesses, 42 hospitalizations, and nine deaths per year in the United States (Scallon *et al.*, 2011). The incidence is low, but the mortality rate is high, and the intensity of treatment and care for survivors is high.

Costs of foodborne illness: US$1,680,903/case in the United States in 2010 (Scharff, 2012).

Testing methods: Culturing techniques for detecting *C. botulinum* in a food sample as well as methods to detect the toxin have been described by Solomon and Lilly (2001) and Maslanka *et al.* (2015). In the United States, any suspected cases of botulism are considered a public health emergency, and involve coordination with the Centers for Disease Control and Prevention for testing done at certified labs.

Controls: The most commonly implicated foods are home-canned, low-acid foods that have been improperly processed. Effective controls to lower the risk of foodborne illness from this species include the following:

- Good sanitation practices will limit the number of spores present in a food.
- Proper heat processing and storage temperatures of hermetically sealed foods (anaerobic conditions). It is important to portion large volumes of foods in shallow containers to facilitate rapid cooling.
- Acidic foods with pH values lower than 4.6 are low risk for illness due to this pathogen.
- Preservatives, such as nitrates, in combination with controlling environmental factors can prevent the growth of *C. botulinum*.

CLOSTRIDIUM PERFRINGENS

Microscopic morphology: This species is a Gram-positive, spore-forming, rod-shaped bacterium. The species name *perfringens* is derived from the Latin word *perfringi*, which means to break or shatter. This species produces a lot of gas when it grows and can burst open wounds or other substrates.

Environmental factors: As with most spore-forming bacteria, these species are ubiquitous in nature and can be found in almost any environment where there is soil and dust. Therefore, it is common to detect low levels of this species in many types of foods. This species can also colonize digestive tracts of animals and can be transmitted through fecal contamination. Humans and animals can be asymptomatic carriers. Foodborne-illness risk increases as conditions become conducive to rapid multiplication of this species in a food system.

This species is a thermoduric mesophile with growth supported by temperatures ranging from 20 to 50°C (68–122°F), and an optimum growth temperature range of 37 to 45°C (98.6–113°F). The spores

Box 6.2. *Clostridium botulinum – blood sausage and pruno*

The word *botulus* is Latin for sausage. The species name for *C. botulinum* is derived from *botulus* due to the implications of blood sausages and botulism.

When botulism was first recognized in Europe, many cases were caused by home-fermented sausages. This derivation, although historically important, has lost much of its significance, since plant rather than animal products are more common vehicles. Sausage now is rarely the cause of botulism in the United States (CDC, 1998).

"After tanking up on 'pruno,' a bootleg prison wine, eight maximum-security inmates at the Utah State prison in Salt Lake County tried to shake off more than just the average hangover" (Wertheim, 2013). After two more similar prison botulism outbreaks investigated by the United States Centers for Disease Control and Prevention (CDC, 2013), prisons were advised to take measures to prevent such outbreaks. These prison outbreaks had a broader impact of using up the antiserum stocks used to treat botulism in that region of the country, putting the rest of the population at high risk of death if other cases occurred at that time. The main source of the toxin in these outbreaks was baked potatoes that had been stolen from the cafeteria and used as part of the mixture to produce pruno. Baked potatoes are an ideal medium for *C. botulinum* growth because the spores are able to survive the initial baking process, and then once in the temperature danger zone and under anaerobic conditions there is ample nutrition and very little competition. The spores can then germinate and vegetative cells grow rapidly and produce toxin. Temperature-abused, foil-wrapped, baked potatoes have been linked to outbreaks of botulism. These prisons were considering removing potatoes and other commonly used carbohydrate sources for pruno from their menus and prison store. Prison inmate and staff trainings were being investigated. This approach would help inmates better understand the risks of pruno consumption.

are heat resistant and may survive cooking, then germinate, and multiply post-heat processing if environmental conditions are favorable. The doubling time is very rapid under optimal temperature conditions and can be as short as 7 minutes.

C. perfringens is a strict anaerobe, and oxygen is toxic to it; however, it has more tolerance than *C. botulinum* due to production of ferredoxin, a reducing molecule that provides some protection from oxygen. *C. perfringens* has been shown to grow under laboratory conditions at a redox potential (E_h) of +320 mV, indicating a tolerance for low concentrations of oxygen.

This species can multiply at pH values higher than 5.0 and requires many nutrients to grow. It does not grow well in the presence of competitive microflora. It also requires a high a_w of about 0.95 to grow.

Type of foodborne illness: This species causes a toxin-mediated infection. Large numbers (over 10^8) of vegetative cells of *C. perfringens* must be ingested to achieve a dosage of the toxin needed to induce symptoms. The toxin, an enterotoxin, is released from the bacterial cell as sporulation is induced. This happens as the vegetative bacterial cell comes into contact with digestive agents of the host. The bacterial cell lyses, releasing the spore and the toxin. The spores can be isolated from fecal samples.

The symptoms that the enterotoxin induces include diarrhea and acute abdominal pain or cramps. The diarrhea is often described as a gassy and frothy diarrhea. The onset time from consumption to symptoms ranges from 6 to 24 hours. The duration of symptoms is usually less than 24 hours. Those who recover from the symptoms may become carriers of *C. perfringens*. See Box 6.3 for examples of outbreaks.

Foodborne illness statistics: *C. perfringens* is estimated to cause 965,958 illnesses, 438 hospitalizations, and 26 deaths per year in the United States (Scallon *et al.*, 2011). Incidence is higher in summer months, and is thought to be due to large gatherings where temperature abuse of foods is more likely to occur. Deaths are rare and are typically due to severe dehydration or necrotic infection of the intestinal tract.

Costs of foodborne illness: US$482/case in the United States in 2010 (Scharff, 2012).

Testing methods: Culturing techniques for detecting and estimating numbers of *C. perfringens* in a

Box 6.3. *Clostridium perfringens* – catering catastrophes

Over 300 people were struck with sickness at a wedding (News Week, 2014). The United States Centers for Disease Control and Prevention confirmed the cause to be *C. perfringens* in the gravy. The caterer was educated on the proper storage of food. This is an unfortunately common scenario.

Erlksen *et al.* (2010) described an outbreak at two different weddings in London that were traced to the same caterer. In these cases, the caterer had not appropriately chilled large volumes of food, allowing for the rapid grow of *C. perfringens*.

Regan *et al.* (1995) described a hospital outbreak of foodborne illness linked to pre-cooked, vacuum-sealed pork. They discussed the implications and the importance for hospitals to review procedures to prevent such outbreaks. Bad catering practices can be particularly traumatic in any patient care environment where victims may already be immunocompromised and in weakened states of health.

In all three of these cases, the foods implicated were meat based and in large quantities. The problem was inadequate cooling post-cooking. The spores of *C. perfringens* can be commonly isolated at low background levels from raw meat. When the meat is cooked in large portions, such as large roasts, or as an ingredient in large volume dishes, the competing microflora are removed and the heat drives out the oxygen. This leads to favorable conditions of nutrients, low competition and anaerobic conditions, and if temperature control is inadequate, the vegetative cells can multiply rapidly. It is very important to cool large volumes of food as rapidly as possible to prevent such microbial activity.

food sample have been described by Rhodehamel and Harmon (2001) and Labbe (2015). This species can be difficult to culture as viability is lost when foods are frozen or held under prolonged refrigeration. Stool samples and epidemiological methods are often better for verifying causation than food samples due to the challenges of viability. Rapid testing methods are also described, as well as toxin detection methods.

Concentrations of *B. cereus* need to be at least 10^5 per gram to implicate it as a causative agent since a high dosage is needed to induce symptoms. Also described are the limitations of these methods. Bennett (2001) describes the methodology for determining the enterotoxigenicity of isolated bacteria.

Controls: Effective controls to lower the risk of foodborne illness from this species include the following:

- Good sanitation practices will limit the number of spores present in a food.
- Proper heat processing of large volumes of food and of hermitically sealed foods (anaerobic conditions). It is important to portion large volumes of foods in shallow containers to facilitate rapid cooling.
- Acidic foods with pH values lower than 4.6 are low risk for illness due to this pathogen.
- Preservatives, such as nitrates, in combination with controlling environmental factors can prevent the growth of *C. perfringens*

The non-spore formers

The following bacterial pathogens of concern for foodborne illness are Gram positive, but do not form spores.

Listeria monocytogenes

Microscopic morphology: This species is a Gram-positive, non-spore-forming, rod-shaped bacterium. The shape is a short rod, also described as a coccal bacillus (see Fig. 6.2). This genus was named to honor Joseph Lister, the pioneer of sterile surgery. The species name, *monocytogenes*, is because of the way this pathogen infects monocytes.

Environmental factors: *L. monocytogenes* is an environmental microorganism and can be found in low numbers almost anywhere lactic acid bacteria can be found. It also colonizes human and animal digestive tracts (asymptomatic carriers) and can be spread through fecal contamination. It is a particularly troublesome problem in food processing plants as it forms very resistant biofilms in areas that have moisture, and can become a post-processing contaminant.

L. monocytogenes is a psychrotroph and can grow in a wide temperature range of 1–45°C (34–114°F), with an optimal growth temperature of 37°C (98.6°F). This pathogen is of most concern in refrigerated ready-to-eat foods

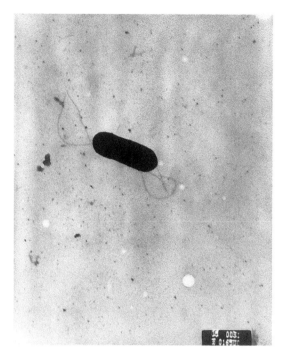

Fig. 6.2. Transmission electron micrograph of a flagellated *Listeria monocytogenes* cell, magnified 41,250× (CDC PHIL, 2014; Balasubr Swaminathan and Peggy Hayes).

transfer can happen across the placenta. The fetus in a pregnant female is at high risk of death from this type of infection. Healthy pregnant women may experience mild, flu-like symptoms, but then experience abortion, premature birth, or stillbirth.

The infection symptoms resemble flu or mononucleosis symptoms. In otherwise healthy adults, the duration of symptoms ranges from one to several weeks. For infants that survive, the duration of symptoms can last 1–4 weeks after birth. In serious cases, meningitis, sepsis, organ abscesses, and death can occur.

See Box 6.4 for an example of outbreaks with public health implications.

Foodborne illness statistics: *L. monocytogenes* is estimated to cause 1,591 illnesses, 1,455 hospitalizations, and 255 deaths per year in the United States (Scallon *et al.*, 2011). Foodborne listeriosis is a rare but serious infectious disease with a high mortality rate of 20–30% (Nyarko and Donnelly, 2015). For the general population, the risk of choking to death on food is higher than of acquiring listeriosis. However, when listeriosis occurs in outbreaks, the severity of the illness tends to garner public outrage. This level of outrage is the reason for the strict regulation in the United States for the presence (zero tolerance) of *L. monocytogenes* in ready-to-eat foods.

Costs of foodborne illness: US$1,282,069/case in the United States in 2010 (Scharff, 2012).

Testing methods: Culturing techniques for detecting and estimating numbers of *L. monocytogenes* in a food sample have been described by Hitchins and Jinneman (2011) and Ryser and Donnelly (2015). *L. monocytogenes* can be difficult to culture and requires special culture-enrichment steps to increase the numbers of cells to detectable levels. Sensitivity and accuracy are important factors when testing for this pathogen. In the United States, any presence of *L. monocytogenes* in a 50g sample of ready-to-eat foods equals adulteration and the food is subject to recall or seizure. The International Commission on Microbiological Specification for Foods (ICMSF) has more liberal specifications not to exceed 100 CFU/g at the point of consumption for individuals not at risk.

For any microbial species that can be a common environmental contaminant, strain typing is an important part of testing when trying to determine

that will not be put through heating prior to eating. As a psychrotroph, it can grow slowly at refrigeration temperatures.

L. monocytogenes can grow at a_w values as low as 0.90 and at pH values of 4.1–9.6, with an optimum pH range of 6.0–8.0. It can survive for weeks in acidic foods (Ryser and Donnelly, 2015). It is a facultative anaerobe, can grow with or without oxygen, and has been a problem in vacuum-packaged foods.

Type of foodborne illness: This species causes an infection. The state of the host is a critical factor for the severity of the infection. Non-pregnant healthy individuals are very resistant to listeriosis. *L. monocytogenes* strains can be very virulent to the immunocompromised, and low cell numbers can cause infection in hosts that are immunocompromised.

The infection process for this pathogen is complex and involves the bacterial cell invading the host's cells. *L. monocytogenes* can invade immune cells, hide and escape from the phagosomes, and be transported to multiple tissues in the host. This

In 2011, a large outbreak that occurred across 28 states within the United States was linked to the consumption of fresh cantaloupe contaminated with *L. monocytogenes*, and resulted in 147 illnesses, 33 fatalities, and one miscarriage (McCollum *et al.*, 2013). The melons were all from the same farm in Colorado. Two years after the outbreak, when brought up on criminal charges, two farmers pleaded guilty to six counts of introducing adulterated cantaloupe to interstate commerce (Andrews, 2014).

> The criminal case against the Jensens has become a milestone in foodborne illness litigation as it was one of the first cases in which food producers faced criminal charges for their contaminated food.... It was the deadliest foodborne illness outbreak recorded in the United States since the CDC began tracking outbreaks in the 1970s (Carter, 2015).

Fresh produce carries many bacteria, including *L. monocytogenes*. Fresh produce that is physically damaged or cut is particularly susceptible to microbial growth as the nutrients of the plant cells are exposed. Fresh, cut produce is kept out of the temperature danger zone for this reason. If temperature control does not also control time, then a psychrotrophic pathogen such as *L. monocytogenes* can grow to infectious dose levels. It is good practice for fresh produce to be thoroughly cleaned and sanitized prior to cutting. In melon production, the melons are cleaned and chilled prior to distribution. In the Jensens' farm cantaloupe outbreak, the brothers had installed new processing equipment, but failed to set up the chlorine spray portion of the process. This control step along with other major sanitation issues resulted in massive cross-contamination of melons. The volume of contaminated ready-to-eat melons, the nature of this pathogen, and the distribution system resulted in the large, high-profile outbreak.

Listeriosis outbreaks have historically been mostly associated with foods of animal origin. Increasingly, produce is becoming a more common vehicle due to modern mass production and supply chain practices.

the source of contamination, particularly for *L. monocytogenes* (Nyarko and Donnelly, 2015).

Controls: Effective controls to lower the risk of foodborne illness from this species include the following:

- Good sanitation practices will limit the number of bacteria present in a food.
- Time is an important control since this pathogen can grow (slowly) at refrigeration temperatures. Limiting time helps to prevent infectious dosage levels from being achieved. FIFO, which stands for first-in first-out, is a food storage rotation strategy to help limit storage time.
- Consistent use of good agricultural and manufacturing practices is important to minimize opportunities for inoculation and cross-contamination.
- Immunocompromised populations and pregnant women are advised to avoid ready-to-eat foods that have a higher risk of *L. monocytogenes* contamination, such as soft cheeses and deli meats. Foods can be heated to 50°C to effectively kill *L. monocytogenes*.
- Environmental testing to ensure effective sanitation is key in preventing cross-contamination with this pathogen (ICMSF, 2002).

Staphylococcus aureus

Microscopic morphology: This species is a Gram-positive, non-spore-forming, coccus-shaped (round) bacterium. *Aureus* was a gold cold of ancient Rome. The genus name, *Staphylococcus*, is derived from *staphyle*, the Greek word for a bunch of grapes and describes how cell clusters appear in a Gram stain under the microscope (see Fig. 6.3).

Environmental factors: *S. aureus* are host-adapted commensal bacteria that are commonly found on skin. The highest concentrations are usually found around moist openings such as the nose, mouth, boils, and sores. *S. aureus* is a common cause of mastitis in cows, and can be a contaminant in milk as a result. Low numbers of this bacteria can be commonly found in foods of animal origin, or those handled by humans (see Box 6.5). Foodborne-illness risk is low as long as the numbers remain low.

S. aureus is a mesophile. It can grow in a temperature range of 7–47.8°C (44–118°F). The optimum temperature range for growth and toxin production is 40–45°C (104–113°F). *S. aureus* is a facultative anaerobe, and can grow with or without oxygen, although it produces less toxin under anaerobic conditions. It requires many nutrients to grow, and does not compete well with other microorganisms. It has a tolerance for salt (sodium chloride). It can grow in a pH range of 4.0–9.8. *S. aureus* is the highest risk for foodborne illness in low-acid foods that are contaminated and temperature abused post-processing.

Type of foodborne illness: This species causes an intoxication. *S. aureus* produces a heat-stable enterotoxin. Large numbers (10^6–10^8) of *S. aureus* are needed to produce enough toxin to induce symptoms. The enterotoxin can survive heating, so even after the bacteria are killed, the toxin remains biologically

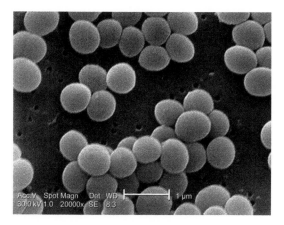

Fig. 6.3. Scanning electron micrograph of *Staphylococcus aureus*, magnified 20,000× (CDC PHIL, 2014; Matthey J. Arduino).

active. Symptoms include profuse vomiting and nausea. Diarrhea can also occur. Severe cases can result in life-threatening dehydration. The onset time from consumption of the toxin to symptom formation ranges from 1 to 6 hours, and the duration ranges from 24 to 48 hours.

Readers should be aware that certain strains of *S. aureus*, such as methicillin-resistant *S. aureus* (MRSA), are high-risk nosocomial (disease originating in a hospital) pathogens. These are of concern for invasive skin infections that can be life threatening and difficult to treat with antibiotics. This form of disease is spread through transmission of dermal bacteria, but is not typically a foodborne illness. However, these strains have been identified in foodborne illness outbreaks due to contamination and temperature abuse of food with subsequent enterotoxin production (Jones *et al.*, 2002; Marler, 2010).

Foodborne illness statistics: *S. aureus* is estimated to cause 241,148 illnesses, 1,064 hospitalizations, and 6 deaths per year in the United States (Scallon *et al.*, 2011). Staphylococcal intoxications are very common, but are mostly unreported due to the self-limiting nature of the illness. Staphylococcal intoxications could be the leading cause of foodborne illness worldwide, but statistics are limited to reported illnesses.

Costs of foodborne illness: US$695/case in the United States in 2010 (Scharff, 2012).

Testing methods: Culturing techniques for detecting and estimating numbers of *S. aureus* in a food

---

Box 6.5. *Staphylococcus aureus* – to glove or not to glove?

Hand hygiene of food handlers is a critical factor in control of foodborne pathogens, especially for *S. aureus*, which is naturally found on skin. Gloves are often used to protect foods from skin contact. This is only effective if the gloves themselves do not become fomites and transfer foodborne pathogens.

Basch *et al.* (2015) documented the frequency of glove changing among food workers who were exchanging money for food in New York City. They indicated that 56.9% of money exchanges observed were not followed up with a glove change. The authors discuss how previous work has indicated that pathogens such as *S. aureus* can survive on and be transferred on money. Other studies have

demonstrated how *S. aureus* can be transferred by gloves (Moore *et al.*, 2013).

Regulations around glove use have the intent to prevent transmission of pathogens such as *S. aureus*. Arguments against glove rules make the case that gloves do not effectively protect against transient bacteria that are the most significant risk of causing foodborne illness, and that the environment of the gloved hand (moist and warm) actually results in more skin bacteria for potential transmission (Snyder, 2001). Gloves may lower but do not eliminate risk of microbial transmission from handler to food (Ronnqvist *et al.*, 2014), and whether gloved or not, effective hand washing is still a critical point of foodborne-illness risk reduction.

---

sample are described by Bennett and Lancette (2001) and Bennett *et al.* (2015a). Foods implicated in foodborne illness cases either have high numbers (~10^6 per gram) of the bacteria (method sensitivity can be low) or, if the bacteria are no longer viable, detection of the enterotoxin (sensitivity needs to be high) can be confirmed.

Controls: Effective controls to lower the risk of foodborne illness from this species include the following:

- Good sanitation practices, especially post-processing, will limit the number of bacteria present in a food. Limiting skin contact with food, especially from open sores or skin infections, prevents inoculation of the food. Effective hand washing practices for food handlers are important to control spread of this and many other contaminants.
- Proper heat processing of large volumes of foods is essential. It is important to portion large volumes of foods in shallow containers to facilitate rapid cooling.
- Acidic foods with pH values lower than 4.6 are lower risk for illness due to this pathogen.

Summary

The difference between classification of bacteria as Gram positive or Gram negative has been explained. Five Gram-positive bacteria of most concern for foodborne illness have been described: *B. cereus, C. botulinum, C. perfringens, L. monocytogenes,* and *S. aureus.* The first three of these are spore-forming bacteria. *B. cereus, C. botulinum* and *S. aureus* cause foodborne intoxications in which toxins are pre-formed in contaminated food. *B. cereus* produces two forms of toxin, an emetic toxin that results in intoxication and a diarrheal toxin that is released during a toxin-mediated infection. *C. perfringens* causes a toxin-mediated infection. *L. monocytogenes* causes an infection and is a psychrotropic bacteria of concern in ready-to-eat foods. Controls for each pathogen have been listed and all include good sanitation practices and temperature control.

Further Reading

FDA (2012) *Bad Bug Book*, 2nd edn. United States Food and Drug Administration, Washington, DC, USA. Available at: http://www.fda.gov/Food/FoodborneIllness Contaminants/CausesOfIllnessBadBugBook/default. htm (accessed 29 December 2015).

Juneja, V.K. and Sofos, J.N. (2010) *Pathogens and Toxins in Foods: Challenges and Interventions.* ASM Press, Washington, DC, USA.

References

Agata, N., Ohta, M., and Yokoyama, K. (2002) Production of *Bacillus cereus* emetic toxin (cereulide) in various foods. *International Journal of Food Microbiology* 73, 23–27.

Andrews, J. (2014) Cantaloupe farmers ask for probation in criminal case. Food Safety News. Available at: http://www.foodsafetynews.com/2014/01/cantaloupe-farmers-ask-for-probation-in-criminal-case/#. Vo1lc03SlaR (accessed 29 December 2015).

Asaeda, G., Bilbert, C., and Swanson, C. (2005) Fried rice syndrome. *Journal of Emergency Medical Services* 30, 30–32.

Basch, C.H., Guerra, L.A., MacDonald, Z., Marte, M., and Bash, C.E. (2015) Glove changing habits in mobile food vendors in New York City. *Journal of Community Health* 40, 699–701.

Bennett, R. (2001) Chapter 15. *Bacillus cereus* diarrheal enterotoxin. In: *Bacteriological Analytical Manual (BAM).* United States Food and Drug Administration, Washington, DC, USA. Available at: http://www.fda.gov/Food/FoodScienceResearch/ LaboratoryMethods/ucm2006949.htm (accessed 29 December 2015).

Bennett, R.W. and Lancette, G.A. (2001) Chapter 12. *Staphylococcus aureus.* In: *Bacteriological Analytical Manual (BAM).* United States Food and Drug Administration, Washington, DC, USA. Available at: http://www.fda.gov/Food/FoodScienceResearch/ LaboratoryMethods/ucm2006949.htm (accessed 29 December 2015).

Bennett, R.W., Hait, J.M., and Tallent, S.M. (2015a) Chapter 39. *Staphylococcus aureus* and staphylococcal enterotoxins. In: *Compendium of Methods for the Microbiological Examination of Foods.* Salfinger, Y. and Tortorello, M.L. (eds) Association of Public Health Laboratories, Washington, DC, USA, pp. 509–526.

Bennett, R.W., Tallent, S.M., and Hait, J.M. (2015b) Chapter 31. *Bacillus cereus* and *Bacillus cereus* toxins. In: Salfinger, Y. and Tortorello, M.L. (eds) *Compendium of Methods for the Microbiological Examination of Foods.* Association of Public Health Laboratories, Washington, DC, USA, pp. 375–390.

Carter, C. (2015) The 2011 Listeria outbreak: outcome of the Jensen criminal trial. Available at: http://foodpolicy.about.com/od/In_the_News/a/Jensen-Brothers-Plead-Guilty-To-Food-Adulteration-Charges.htm (accessed 29 December 2015).

CDC (1998) *Botulism in the United States, 1899–1996. Handbook for Epidemiologists, Clinicians, and Laboratory*

Workers. Centers for Disease Control and Prevention, Atlanta, GA, USA.

CDC (2013) Notes from the field: botulism from drinking prison-made illicit alcohol – Arizona, 2012. *Morbidity and Mortality Weekly Report.* Centers for Disease Control and Prevention, Atlanta, GA, USA.

CDC PHIL (2014) United States Centers for Disease Control Public Health Image Library. Available at: http://phil.cdc.gov/phil/home.asp (accessed 29 December 2015).

Erlksen, J., Zenner, D., Anderson, S.R., Grant, K., and Kumar, D. (2010) *Clostridium perfringens* in London, July 2009: two weddings and an outbreak. *Euro Surveillance* 15, 1–6.

Hitchins, A.D. and Jinneman, K. (2011) Chapter 10. Detection and enumeration of *Listeria monocytogenes.* In: *Bacteriological Analytical Manual (BAM).* United States Food and Drug Administration, Washington, DC, USA. Available at: http://www.fda.gov/Food/FoodScienceResearch/LaboratoryMethods/ucm2006949.htm (accessed 29 December 2015).

ICMSF (2002) *Microorganisms in Foods 7: Microbiological Testing in Food Safety Management.* International Commission on Microbiological Specifications for Foods. Kluwer Academic/Plenum, New York, USA.

Jones, T.F., Kellum, M.E., Porter, S.S., Bell, M., and Schaffner, W. (2002) An outbreak of community-acquired foodborne illness caused by methicillin-resistant *Staphylococcus aureus. Emerging Infectious Diseases* 8, 82–84.

Labbe, R.G. (2015) Chapter 33. *Clostridium perfringens.* In: Salfinger, Y. and Tortorello, M.L. (eds) *Compendium of Methods for the Microbiological Examination of Foods.* Association of Public Health Laboratories, Washington, DC, USA, pp. 403–409.

Marler, B. (2010) About MRSA (methicillin-resistant *Staphylococcus aureus*). *Food Poison Journal.* Available at: http://www.foodpoisonjournal.com/food-poisoning-information/about-mrsa-methicillinre-sistant-staphylococcus-aureus/#.Vo5wiE3SIaQ (accessed 29 December 2015).

Maslanka, S.E., Solomon, H.M., Sharma, S., and Johnson, E.A. (2015) Chapter 32. *Clostridium botulinum* and its toxins. In: Salfinger, Y. and Tortorello, M.L. (eds) *Compendium of Methods for the Microbiological Examination of Foods.* Association of Public Health Laboratories, Washington, DC, USA, pp. 391–401.

McCollum, J.T., Cronquist, A.B., Silk, B.J., Jackson, K.A., O'Connor, K.A., Cosgrove, S., Gossack, J.P., Parachini, S.S., Jain, N.S., Ettestad, P., Ibraheem, M., Cantu, V., Joshi, M., DuVernoy, T., Fogg, N.W. Jr, Gorny, J.R., Mogen, K.M., Spires, C., Teitell, P., Joseph, L.A., Tarr, C.L., Imanishi, M., Neil, K.P., Tauxe, R.V., and Mahon, B.E. (2013) Multistate outbreak of listeriosis associated with cantaloupe. *New England Journal of Medicine* 369, 944–953.

Moore, G., Dunnill, C.W., and Wilson, P.R. (2013) The effect of glove material upon the transfer of methicillin-resistant *Staphylococcus aureus* to and from a gloved hand. *American Journal of Infection Control* 41, 19–23.

News Week (2014) More than 300 sickened at Missouri wedding. *News Week Magazine.* Available at: http://www.foodsafetynews.com/2014/05/300-sickened-at-mo-wedding/#.Vo0w103SIaR (accessed 29 December 2015).

Nyarko, E.B. and Donnelly, C.W. (2015) *Listeria monocytogenes*: strain heterogeneity, methods, and challenges of subtyping. *Journal of Food Science* 80, M2868–M2878.

Regan, C.M., Syed, Q., and Tunstall, P.J. (1995) A hospital outbreak of *Clostridium perfringens* food poisoning – implications for food hygiene review in hospitals. *Journal of Hospital Infections* 29, 69–73.

Rhodehamel, E.J. and Harmon, S.M. (2001) Chapter 16. *Clostridium perfringens.* In: *Bacteriological Analytical Manual (BAM).* United States Food and Drug Administration. Washington, DC, USA. Available at: http://www.fda.gov/Food/FoodScienceResearch/LaboratoryMethods/ucm2006949.htm (accessed 29 December 2015).

Ronnqvist, M., Aho, E., Mikkela, A., Ranta, J., Tuomlnen, P., Ratto, M., and Maunula, L. (2014) Norovirus transmission between hands, gloves, utensils, and fresh produce during simulated food handling. *Applied and Environmental Microbiology* 80, 5403–5410.

Ryser, E.T. and Donnelly, C.W. (2015) Chapter 35. *Listeria.* In: Salfinger, Y. and Tortorello, M.L. (eds) *Compendium of Methods for the Microbiological Examination of Foods.* Association of Public Health Laboratories, Washington, DC, USA, pp. 425–443.

Scallon, E., Hoekstra, R.M., Angulo, F.J., Tauxe, R.V., Widdowson, M.A., Roy, S.L., Jones, J.L., and Griffin, P.M. (2011) Foodborne illness acquired in the United States – Major Pathogens. *Emerging Infectious Diseases* 17, 7–15.

Scharff, R.L. (2012) Economic burden from health losses due to foodborne illness in the United States. *Journal of Food Protection* 75, 123–131.

Setlow, P. and Johnson, E.A. (2013) Spores and their significance. In: Doyle, M.P. and Buchanan, R.L. (eds) *Food Microbiology: Fundamentals and Frontiers,* 4th edn. ASM Press, Washington, DC, USA, pp. 45–79.

Snyder, O.P. (2001) Why gloves are not the solution to the fingertip washing problem. *Hospitality Institute of Technology and Management.* Available at: http://w.highfield.co.uk/download/sofs/Glove-problems.pdf (accessed 29 December 2015).

Solomon, H.M. and Lilly, T. (2001) Chapter 17. *Clostridium botulinum.* In: *Bacteriological Analytical Manual (BAM).* United States Food and Drug Administration, Washington, DC, USA. Available at: http://www.fda.gov/Food/FoodScienceResearch/LaboratoryMethods/ucm2006949.htm (accessed 29 December 2015).

Tallent, S.M., Rhodehamel, E.J., Harmon, S.M., and Bennett, R. (2012) Chapter 14. *Bacillus cereus*. In: *Bacteriological Analytical Manual (BAM).* United States Food and Drug Administration, Washington, DC, USA. Available at: http://www.fda.gov/Food/FoodScienceResearch/Laboratory Methods/ucm2006949.htm (accessed 29 December 2015).

Wertheim, B. (2013) How not to die of botulism. *The Atlantic.* Available at: http://www.theatlantic.com/health/archive/2013/12/how-not-to-die-of-botulism/281649/ (accessed 29 December 2015).

7 Gram-Negative Bacteria

Key Questions
- Which Gram-negative bacteria are of most concern for microbial food safety?
- What are the mechanisms by which these Gram-negative bacteria cause illness?
- What are the hazards that these Gram-negative bacteria present to consumers?
- What control measures are available to prevent foodborne illness due to these Gram-negative bacteria?

In Chapter 6, we explained the difference between Gram-positive and Gram-negative bacteria. In this chapter, we focus on the Gram-negative bacteria of most concern for foodborne illness. One major distinction is that none of these Gram-negative bacteria causes intoxications, and another is that none of the Gram-negative bacteria is a spore former. The bacteria discussed in this chapter are *Campylobacter*, *Escherichia coli*, *Salmonella*, *Shigella*, *Vibrio*, and *Yersinia*.

Campylobacter

Microscopic morphology: This genus of Gram-negative bacteria comprises rod shaped, small cells that are corkscrew-shaped or spirally curved. The genus name was derived from the Greek word for curved rod. These bacteria are motile, moving with the aid of flagella and their curved shapes (see Fig. 7.1). There are over 20 species within the genus, and the two of these most commonly associated with foodborne illness outbreaks are *C. jejuni* and *C. coli*.

Environmental factors: *Campylobacter* spp. are challenging to culture due to the specificity of the environmental factors needed to meet the requirements for growth. These bacteria are typically associated with warm-blooded animals, and are commonly found in poultry. Infected dairy cows can be a source of transmission through raw milk (see Box 7.1). Almost any food animal and humans can be carriers of *Campylobacter*.

Campylobacter require warm temperatures for growth, ranging from 30 to 45°C (86–113°F), and an optimum growth temperature around 42°C (107.6°F). This is a narrow range of temperatures in comparison to other foodborne pathogens. *Campylobacter* are sometimes classified as thermophiles, although most would technically be on the high end of the temperature range for mesophiles (see Chapter 3). The optimal growth temperature would be close to the normal body temperatures of poultry, which tend to have higher body temperatures than other larger animal livestock.

Campylobacter are classified as microaerophilic and require a low-oxygen atmosphere. Normal oxygen levels in the atmosphere are around 21%, which is toxic to *Campylobacter*. These species will only grow at 3–6% oxygen. They are not anaerobic, as some oxygen is required. They also require 10% carbon dioxide.

These bacteria can tolerate sodium chloride (table salt) concentrations as high as 3.5%, so can survive as post-process contaminates in certain processed foods. They do not survive cooking, are sensitive to freezing, and are sensitive to non-neutral pH. The prevalence of campylobacteriosis is quite remarkable considering the environmental limitations for survival and growth.

Type of foodborne illness: *C. jejuni* and *C. coli* cause intestinal tract infections. The mechanisms of the infection process are not clear, and there may be toxins involved. Regardless of the mechanism, the result is damage to the intestinal lining, causing diarrhea, which can be bloody in severe cases. The damage causes abdominal pain or cramps similar to an appendicitis. Other symptoms include malaise, headache, and fever consistent with an infection.

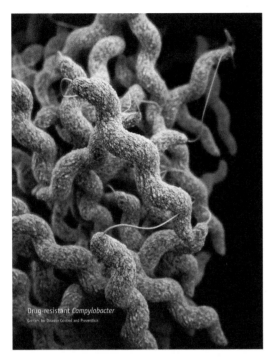

Drug-resistant *Campylobacter*
Centers for Disease Control and Prevention

Fig. 7.1. Three-dimensional computer-generated image of a cluster of *Campylobacter* (CDC PHIL, 2014; James Archer).

Virulent strains may have infectious doses as low as 500 cells, so not much contamination is required to raise the risk of foodborne illness. The onset time for symptoms ranges from 2 to 10 days, and the symptoms can recur for up to several weeks. Those who recover may shed these bacteria for months after symptoms clear.

In some cases, a post-infectious arthritis can occur. This is called reactive arthritis (Ajene *et al.*, 2013). Rates of occurrence are not yet well tracked as diagnosis is challenging, but reactive arthritis may affect about 8% of adults and 6% of children after a case of campylobacteriosis. Symptoms may develop weeks to months after the infection clears, further contributing to the morbidity of campylobacteriosis. Some forms may resolve in about 6 months, but about 63% of cases remain chronic.

In very rare cases, a complication known as Guillain–Barré syndrome can occur. This syndrome is the result of an autoimmune response that damages nerves and leaves the host paralyzed. This paralysis is usually reversible, but about 20% of cases result in disability and 5% die (Yuki and Hartung, 2012).

Foodborne illness statistics: The species of *C. jejuni* and *C. coli* are difficult to distinguish, and most foodborne illness statistics do not distinguish between them. *Campylobacter* are estimated to cause 845,024 illnesses, 8,463 hospitalizations, and 76 deaths in the United States per year (Scallon *et al.*, 2011).

Costs of foodborne illness: *Campylobacter* foodborne infections were estimated to cost US$8,141/case in the United States in 2010 (Scharff, 2012).

Box 7.1. *Campylobacter* – poultry and raw milk

The major causes of outbreaks of campylobacteriosis are raw poultry and raw milk (Taylor *et al.*, 2012).

Much effort has gone into preventative control strategies in the poultry industry, and this has resulted in a decline of outbreaks linked to poultry. With the decline of poultry contamination, notable increases in other foods have occurred, including in produce. This is likely due to contaminated water supplies used for production and processing. Careful management of poultry production, including manure management, remains an area for attention.

Raw milk has become the most common cause of campylobacteriosis in the United States, and outbreaks spike where raw milk producers market these risky products. Outbreaks tend to cluster geographically around raw milk producers in states where raw milk sales are still legal, but as they cannot sell raw milk and related products over state lines, this makes finding the source of the cause easier. These types of outbreaks are of particular concern for immunocompromised people such as small children, for whom the risk of death or permanent disability is higher. Raw milk enthusiasts continue to resist regulatory measures to prevent such outbreaks, even though, on balance, the clear health risks far outweigh the benefits promoted for raw milk.

Chapter 7

Testing methods: Standard testing protocols for *Campylobacter* in foods have been described by Hunt *et al.* (2001) and Corry and Line (2015). Sensitivity is important to detect low levels in foods and beverages, and culture methods include enrichment steps to raise the number of bacteria to detectable levels. It is important that samples are not put in frozen storage prior to analysis, as freezing will damage the viability of these bacteria.

Controls: Effective controls to lower the risk of foodborne illness from this species include the following:

- Good agricultural, sanitation, and hygiene practices will limit the number of bacteria present in a food.
- Temperature control that keeps food out of the temperature danger zone (4–60°C; 40–140°F) is important to prevent growth.
- Avoiding cross-contamination is important, especially between raw meat and ready-to-eat foods. Education of food handlers, including consumers, is important to ensure an understanding of how to prevent cross-contamination.
- Pasteurizing milk is the most effective control to prevent transmission of *Campylobacter* through milk.

Escherichia coli

Microscopic morphology: This species of Gram-negative bacteria is rod shaped (see Fig. 7.2). The genus was named after the first person who isolated this organism, the German pediatrician Theodor Escherich. The species name, *coli*, was derived from the word colon, which is where this species can commonly be isolated from.

Environmental factors: *E. coli* is a prominent gastro-intestinal microorganism. This species includes a large number of strains that represent the broad genetic diversity of this species. Most strains are non-pathogenic commensals and are important for digestive health. Rarely, a strain may pick up some genes that give new virulence capabilities, and it then become an opportunistic pathogen. The gastro-intestinal tract is an environment with high concentrations of many microorganisms, and mechanisms of horizontal gene transfer can lead to the formation of new genetic strains. In *E. coli*, genetic transformations may result in different serotypes and virotypes.

Antigens in their outer cell components are used to designate different strains of *E. coli*. This is called serotyping. The two main types of antigens

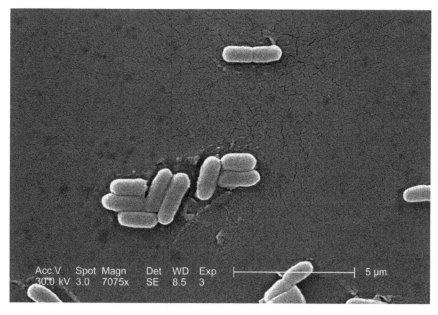

Fig. 7.2. Scanning electron micrograph of *Escherichia coli*, magnified 7,075× (CDC PHIL, 2014; Janice Carr). Note that, morphologically, *E. coli*, *Salmonella* and *Shigella* look similar under the microscope.

are designated O, for outer membrane antigens, and H for flagellar antigens. There are around 200 types of O antigens and over 30 types of H antigens. One serotype of *E. coli* that is pathogenic is O157:H7, where the numbers identify the specific O and H antigens.

Virotypes, also termed pathotypes or pathovars (Croxen and Finlay, 2010), refer to differentiation of pathogenic strains based on pathogenic virulence properties. The United States Centers for Disease Control and Prevention describes six pathotypes of *E. coli* (CDC, 2015a). The scientific literature can vary on the terminology used for virotypes. There are six common virotypes used for pathogenic *E. coli*:

- enteroaggregative *E. coli* (EAggEC);
- enterohemorrhagic *E. coli* (EHEC), also referred to as shiga toxin-producing (STEC) or verocytotoxin-producing (VTEC);
- enteroinvasive *E. coli* (EIEC);
- enteropathogenic *E. coli* (EPEC);
- enterotoxigenic *E. coli* (ETEC); and
- diffusely adherent *E. coil* (DAEC).

All strains of *E. coli* can be transmitted through food, and the presence of any *E. coli* in food is an indicator of fecal contamination. *E. coli* has been a common problem in foods of animal origin, but is increasingly being linked to foodborne illness outbreaks from foods of plant origin (see Box 7.2). *E. coli* can survive outside of animal hosts for long periods of time, and can be easily transmitted through fertilizers, soil, and water.

Temperature growth ranges can vary among strains due to their genetic diversity. For strains of O157:H7, temperature growth ranges are from 8 to 42°C (46.4–107.6°F) with an optimum growth temperature of 37°C (98.6°F; human body temperature). Pathogenic strains have more acid tolerance and can survive at pH 3.6. This has led to problems in fruit juice production, and in the United States, fruit juices are required to be pasteurized for this reason.

Type of foodborne illness: The type of illness depends on the virotype.

The EAggEC strains cause a toxin-mediated infection. They will attach to intestinal microvilli and then excrete hemolysis-like toxins that induce a persistent diarrhea. This syndrome is more common in children where the diarrhea can last over 2 weeks.

The EHEC strains cause a toxin-mediated infection. These strains are a subgroup of STEC strains, are most frequently implicated in severe clinical illness worldwide, and include seven priority serotypes: O26:H11, O45:H2, O103:H2, O111:H8, O121:H19, O145:H28, and O157:H7 (Delannoy *et al.*, 2013). They will attach to intestinal microvilli and excrete shiga-like toxins. Shiga-like means that there are similarities to toxins produced by *Shigella* spp. EHEC shiga-like toxins are enterotoxic and

---

Box 7.2. *Escherichia coli* – burgers and sprouts

When hemorrhagic strains of *E. coli* first started to emerge as foodborne pathogens, the most common food implicated was ground beef. Large outbreaks were commonly traced back to processors, as in the 2002 multistate outbreak in the United States linked to ground beef products that led to a record amount of ground beef being recalled (CDC, 2002). This recall involved 354,200 pounds of meat. These types of outbreaks led to much research into the proper ways to handle and cook ground beef. This research concluded that cooking to an internal temperature was a much more effective control compared to traditional methods of determining doneness of cooking such as color (USDA, 2013). You can see the current recommended cooking temperatures at FSIS (2015).

Although raw meat remains an important source of pathogenic *E. coli*, outbreaks have seemed to shift more to raw produce. A very large *E. coli* O104:H4 outbreak occurred in Europe in 2011 (Rasko *et al.*, 2011). This outbreak included over 4,000 cases of illness across 16 countries (CDC, 2013). The European Union banned the importation of fenugreek and various other seeds, including beans and sprouts, from Egypt, believing them to be the source of the outbreak. Raw seed sprouts, a food considered highly nutritious, have been a notorious transmission source for enteric pathogens such as *E. coli* (Malbury, 2015). The seeds serve as a nutrient source for bacteria, which can flourish during the high-moisture sprouting process. Research continues to find ways to better control the risk level of foodborne illness transmission by raw sprouts.

---

cause diarrhea. They are also cytotoxic and damage host cells, which can result in bloody diarrhea. Symptoms of diarrhea develop within 4 days of ingestion, and for those that recover, the duration is typically 3 to 7 days. These toxins can enter the blood stream and cause severe kidney damage. This type of infection can become life threatening, especially if kidney damage is severe. Antibiotics only aggravate the symptoms as they induce increased toxin production by the pathogen. It is estimated that fewer than ten cells of a virulent strain of *E. coli* can constitute and infectious dose.

The EIEC strains cause an infection. The cells invade intestinal tract host cells and cause a voluminous bloody diarrhea. The onset time varies from 2 to 48 hours. This is a short onset period for an infection, and reflects the invasiveness of this pathogenic strain.

The EPEC strains cause an infection. The cells adhere to microvilli in the intestinal tract and can invade the intestinal cells. This infection process results in diarrhea. The EPEC are most commonly associated with pediatric diarrhea in infants.

The ETEC strains cause a toxin-mediated infection. They attach to intestinal microvilli and excrete enterotoxins, one of which is similar to cholera toxin. Symptoms include diarrhea. These strains are the most common cause of traveler's diarrhea (Connor, 2015). Humans are the main reservoir for this type, and human fecal contamination of food or water is the most common cause.

The DAEC strains may cause a toxin-mediated infection. This strain causes a watery diarrhea. The pathogenic mechanisms used by this strain are not yet definitively defined. The categorization is based on the observed patterns of how these strains bind to intestinal cells. This strain is not typically associated with foodborne illness outbreaks.

Foodborne illness statistics: Overall, foodborne pathogenic *E. coli* are estimated to cause 205,781 illnesses, 2,249 hospitalizations, and 20 deaths in the United States per year (Scallon *et al.*, 2011). For just the O157:H7 strains, the estimates are 63,153 illnesses, 2,138 hospitalizations, and 20 deaths. Based on these statistics, the O157:H7 strains clearly cause the most life-threatening forms of *E. coli* foodborne illness in the United States.

Costs of foodborne illness: Overall, foodborne pathogenic *E. coli* were estimated to cost US$14,083/case in the United States in 2010, and of that,

US$10,048/case for those caused by O157:H7 strains (Scharff, 2012).

Testing methods: Methods for testing for *E. coli* have been described by others (Feng *et al.*, 2013, 2015; Kornacki *et al.*, 2015; Meng *et al.*, 2015). Methods range from broad tests for bacteria that fall into the fecal indicator category, to those with specificity that target individual pathogenic serotypes. Fecal indicators are used to check for such contamination in food, and detect more than just *E. coli*. Tests more specific for *E. coli* are more accurate indicators of fecal contamination, but not all strains detected will be pathogenic strains.

When testing for pathogenic strains of *E. coli*, sensitivity of a testing method is very important. In the United States, one cell of *E. coli* O157:H7 detected in 25 g of ground beef is considered adulteration. A detection limit that low can be very challenging to meet. When testing is intended to detect pathogenic strains of *E. coli*, it is very important to ensure that the test is for this purpose. Several standard testing methods for generic *E. coli* can miss detection of virotypes due to differences in enzyme production among strains.

Controls: Effective controls to lower the risk of foodborne illness from this species include the following:

- Good agricultural, sanitation, and hygiene practices will limit the number of bacteria present in a food.
- Temperature control that keeps food out of the temperature danger zone (4–60°C; 40–140°F) is important to prevent growth.
- Avoiding cross-contamination is important, especially between raw meat and ready-to-eat foods. Education of food handlers, including consumers, is important to ensure an understanding of how to prevent cross-contamination.

Salmonella

Microscopic morphology: *Salmonella* spp. are Gram-negative bacteria with a rod shape and several flagella that provide motility. This genus name was derived from the name of one of the discoverers (Salmon and Smith, 1885).

Environmental factors: *Salmonella* are facultative anaerobes and can multiply with or without oxygen present. They can grow at pH ranges of

4 to 9, with an optimum pH of 7. The temperature range for growth of *Salmonella* is 5.3–45°C (41.5–113°F), with an optimum temperature for growth of 37°C (98.6°F).

Salmonella are commonly found in almost any animal's digestive tract, and this leads to its common occurrence in the environment due to fecal contamination (see Box 7.3 for examples of large outbreaks).

Salmonella do not form spores, but can achieve a state of dormancy that allows survival at water activities as low as 0.43. Outbreaks of salmonellosis are increasingly commonly linked to low-moisture foods such as nuts, seeds, and confections like chocolate. The United States Food and Drug Administration recently analyzed data and is looking for input for improving the safety of dried spices, with *Salmonella* contamination being a major hazard of concern (FDA, 2016).

Type of foodborne illness: Virulent strains of *Salmonella* cause infections termed salmonellosis (CDC, 2015b). The two types of salmonellosis are gastroenteritis and enteric fever. The gastroenteritis can be life threatening in immunocompromised individuals, but is typically self-limiting in healthy individuals. Enteric fever, such as typhoid fever, can be life threatening if not treated with antibiotics. Enteric fever is endemic in developing countries and is typically spread due to poor sanitation practices in areas that struggle with clean water supplies. Gastroenteritis is the more common form of salmonellosis in developed countries, but occurs worldwide.

The disease form is linked to the virulence genes in the *Salmonella* strain causing the infection. There is a wide variety of genetic diversity within the genus *Salmonella*. There are technically only two species, *Salmonella enterica* (within which there are over 2,500 serovars) and *Salmonella bongori* (within which there are about 20 serovars), and the number of serovars changes with some frequency over time (Brenner *et al.*, 2000). Of these two species, *S. enterica* causes the vast majority of illnesses in humans. The subtypes are indicated in the literature with the serotype name capitalized, but not italicized. *Salmonella* serotype Typhimurium and *Salmonella* serotype Enteritidis are the most common in the United States (CDC, 2015b), and cause gastroenteritis. Other serotype designations are commonly named after the city of the original outbreak from which a strain has been isolated, i.e. *S.* Montevideo, *S.* London, *S.* Miami, and *S.* Richmond.

S. enterica serovars can be divided into typhoidal and non-typhoidal groups, each of which elicits different types of immune responses and symptoms (Gal-Mor *et al.*, 2014). Typhoid fever is a severe form of enteric fever and is caused by *S.* Typhi, *S.* Sendai, and *S.* Paratyphi. These typhoidal serotypes are very host adapted to humans. Both typhoidal and non-typhoidal *Salmonella* initialize infection by invading the small intestine epithelial cell layer and can invade immune cells (see Fig. 7.3), which transfer them to other organ systems and tissues. The initial immune response in the intestinal tract to the typhoidal strains is mild and results in less gastrointestinal damage compared to non-typhoidal strains. However, the typhoidal strains cause more systemic damage. In most cases,

---

Box 7.3. *Salmonella* – big outbreaks and peppers

In Chapter 6, we described the largest ever recorded interstate outbreak of foodborne illness in the United States, and the cause was *Listeria*. Prior to that outbreak, *Salmonella* was the more common source of large, interstate outbreaks. Some large-scale salmonellosis outbreaks include:

- peanut butter: multistate outbreak of *Salmonella* Typhimurium infections linked to peanut butter in 2008–2009 (CDC, 2009);
- breakfast cereal: multistate outbreak of *Salmonella* serotype Agona infections linked to toasted oats cereal in April and May 1998 (CDC, 1998);

- ice cream: over 200,000 cases, a national outbreak of *Salmonella enteritidis* infections from ice cream (Hennessy *et al.*, 1996).

One outbreak of particular notoriety crossed international borders and involved peppers. In 2008, an outbreak of *Salmonella* Saintpaul infections swept the United States. It crossed 43 states and included 1,500 cases of which 21% were hospitalized and two died (Behravesh *et al.*, 2011; CDC, 2008). The outbreak was initially linked to raw tomatoes, but later peppers emerged as the source. Jalapeño peppers collected in Texas and serrano peppers on a Mexican farm were linked to the outbreak strain.

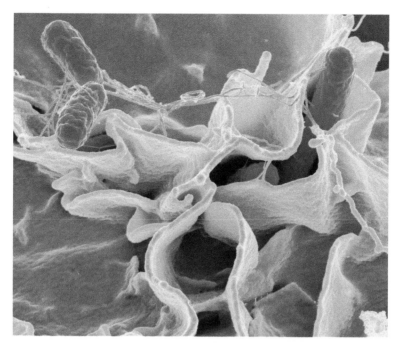

Fig. 7.3. A microscopic view of *Salmonella* bacteria invading an immune cell (*Salmonella* image from Flickr.com through the Creative Commons courtesy of the United States National Institutes of Health (NIH) Image Gallery).

the non-typhoidal strains simply do not travel as far and symptoms are limited to damage in the gastrointestinal tract. Typhoidal strains are more likely to induce a carrier state (see the story of Typhoid Mary in Chapter 5, Box 5.2). Antibiotics are necessary to treat typhoidal strains, but antibiotics can actually exacerbate non-typhoidal salmonellosis. In severe cases of gastroenteritis, where antibiotic treatment is needed, non-typhoidal strains that have broad antibiotic resistance, such as *S.* Typhimurium definitive type 104 (DT 104) continue to emerge, threatening survival for patients with limited to no treatment options (Chen *et al.*, 2013).

Symptoms of typhoid fever include loss of appetite, headache, and high fever. The onset time ranges from 7 to 21 days. The duration can be 3 to 6 weeks without treatment, and the mortality rate is high if untreated. Treatment regimens typically include administration of antibiotics over a 4-week period.

Symptoms of gastroenteritis include nausea, vomiting, abdominal pain, headache, drowsiness, chills, slight fever, and diarrhea. The onset time is typically 12 to 14 hours, and the duration in self-limiting cases is 2 to 3 days. The infectious dose ranges widely from very few cells to 10^8 to establish infection, since virulence characteristics vary so broadly.

As with *Campylobacter* infections, in some cases of salmonellosis a post-infectious reactive arthritis can occur (Ajene, 2013). Rates of occurrence are not yet well tracked as diagnosis is challenging, but reactive arthritis may affect about 12% of adults and 8% of children after a case of salmonellosis. Symptoms may develop weeks to months after the infection clears, further contributing to the morbidity of salmonellosis. Some forms may resolve in about 6 months, but about 63% of cases remain chronic.

Foodborne illness statistics: Typhoidal *Salmonella* are estimated to cause 1,821 illnesses, 197 hospitalizations, and zero deaths per year in the United States per year (Scallon *et al.*, 2011). Non-typhoidal *Salmonella* are estimated to cause 1,027,561 illnesses, 19,336 hospitalizations, and 378 deaths per year in the United States (Scallon *et al.*, 2011). Of confirmed cases of foodborne illnesses, salmonellosis is the second most common cause after norovirus in the United States. The estimated global impact for typhoidal *Salmonella* is 22 million cases of typhoid fever and 200,000 deaths globally each year (Newton *et al.*, 2015). The estimated global impact for non-typhoidal *Salmonella* is 94 million cases of gastroenteritis and 115,000 deaths globally each year (Iwamoto, 2015).

Costs of foodborne illness: Overall, the estimated costs were US$11,488/case in the United States in 2010 for typhoidal *Salmonella* and US$11,086/case in the United States in 2010 for non-typhoidal *Salmonella* (Scharff, 2012).

Testing methods: Methods for detecting and quantifying *Salmonella* in a food have been described by Andrews *et al.* (2015) and Cox *et al.* (2015). Sensitivity is important for testing most foods since numbers of *Salmonella* cells may be low. Enrichment steps to increase numbers of cells to detectable levels are often needed. Serotyping and molecular methods are necessary to identify strain type.

Controls: The most commonly implicated foods are foods from animal origin and fresh produce. Effective controls to lower the risk of foodborne illness from this species include the following:

- Good agricultural, sanitation, and hygiene practices will limit the number of bacteria present in a food.

- Temperature control that keeps food out of the temperature danger zone (4–60°C; 40–140°F) is important to prevent growth.
- Acidic foods with pH values lower than 4.6 are low risk for illness due to this pathogen.
- *Salmonella* are sensitive to high salt concentrations, ionizing radiation, and cooking.

Shigella

Microscopic morphology: *Shigella* spp. are Gram-negative bacteria with a rod shape. The genus name comes for the discoverer, Kiyoshi Shiga (Eiko, 2002). *Shigella* are closely related to *E. coli* and *Salmonella,* and their appearance is similar under the microscope.

Environmental factors: *Shigella* are intestinal bacteria adapted to primates. Their presence in environmental or food samples is typically due to fecal contamination (see Box 7.4).

Shigella are facultative anaerobes and can multiply with or without oxygen present. Their optimum pH range is 6.5 to 7.5, but growth has been demonstrated under laboratory conditions in a pH range of 4 to 9.0.

Box 7.4. *Shigella* – keep poop out of food

Nygren *et al.* (2015) reported on the outbreaks of shigellosis in the United States from 1998 to 2008. These included 120 outbreaks involving 6,208 cases of illness. Most of these were restaurant-associated infections. Raw foods and those contaminated by food handlers accounted for many of these cases. In simplest terms, these outbreaks were due to human excrement being transmitted to the food. Since *Shigella* has a small infectious dose, it does not take much fecal contamination to cause illness. Campaigns to educate food handlers about personal hygiene have been a major component of food safety control measures to prevent enteric diseases such as shigellosis, and rates of shigellosis outbreaks in the United States seem to be on the decline as a result.

One large shigellosis outbreak with 886 cases of illness involved raw tomatoes (Reller *et al.*, 2006). This outbreak included five local restaurants in the state of New York. The investigation indicated that 16 workers were infected with the same *Shigella flexneri* strain as was implicated in the outbreak; two were ill

and 14 were asymptomatic. In this outbreak, the tomatoes implicated were overripe (less acidic) and had been stored in the temperature danger zone. *Shigella* was thought to have entered the tomatoes through the stem scar of damaged tomatoes. The tomatoes were unwashed and had been hand-sorted by workers with uncertain health status at the time. Fecal material had been widely spread through contaminated, raw tomatoes. The authors emphasized that handwashing must occur both at distribution sites and at sites of food preparation.

In 2015, the ten largest United States foodborne illness outbreaks (Zuraw, 2015) ranged in case size from 52 to 838, with nine out of ten outbreaks caused by enteric pathogens, likely spread by fecal contamination. One of these outbreaks was of shigellosis, and included 194 cases of illness. These were all people who had eaten at one seafood restaurant. The exact transmission source remains a mystery; however, one of the employees tested was positive for *Shigella*, but for a different strain from that implicated in the outbreak (Janet O, 2015).

Shigella are considered a mesophile with a temperature growth range of 10–48°C (50–118.4°F), and an optimum growth temperature around 37°C (98.6°F).

Type of foodborne illness: *Shigella* cause a toxin-mediated infection, producing potent cytotoxins called shiga toxins. Shiga toxins help the pathogen to invade host cells. The bacteria can multiply in the host cells, further spreading the infection and severely damaging the intestinal tract. The epithelial layer of the colon is the area where these pathogens do the most host damage, resulting in damage that disrupts the host's ability to retain fluids, and presenting as dysentery. Hosts who recover can become carriers. Dysentery is actually different from diarrhea, and *Shigella* can cause both:

- diarrhea: watery discharge, mainly from the small intestine (the runs); and
- dysentery: frequent but smaller-volume stools than diarrhea that contain blood and/or pus from mucosal damage (the squirts).

There are four species of *Shigella*: *S. dysenteriae*, *S. flexneri*, *S. boydii*, and *S. sonnei*. *S. dysenteriae* causes classical bacillary dysentery in humans, and is extremely virulent with as few as ten cells being enough for an infectious dose. These species are very host adapted to humans, so contamination of water and food will have originated from an infected human. Dysentery remains a serious public health challenge in both developed and developing countries, especially after disruptions to sanitary conditions occur. Polluted water and infected food handlers are of primary concern. Because growth in a contaminated food does not necessarily need to happen to achieve an infectious dose, control strategies can be challenging.

The symptoms of shigellosis have an onset time ranging from 8 to 50 hours. These include watery diarrhea (may contain blood and mucous), cramps, and sometimes fever and vomiting. The duration in an otherwise healthy host is typically 5 to 6 days, and for others can last for weeks. The mortality rate is low if there is treatment. Children are at highest risk of death due to severe dehydration (less overall fluid in a smaller body).

As with *Campylobacter* and *Salmonella* infections, in some cases of shigellosis, a post-infectious reactive arthritis can occur (Ajene, 2013). Rates of occurrence are not yet well tracked as diagnosis is challenging, but reactive arthritis may affect about 12% of adults and 7% of children after a case of shigellosis. Symptoms may develop weeks to months after the infection clears, further contributing to the morbidity of shigellosis. Some forms may resolve in about 6 months, but about 63% of cases remain chronic.

Foodborne illness statistics: *Shigella* is estimated to cause 131,254 illnesses, 1,456 hospitalizations, and ten deaths per year in the United States (Scallon *et al.*, 2011). The estimated global impact for *Shigella* is 80–165 million cases of illness and 600,000 deaths globally each year (Bowen, 2015).

Costs of foodborne illness: Overall, the estimated cost was $9,551/case in the United States in 2010 (Scharff, 2012).

Testing methods: Because of similarities to *E. coli* and *Salmonella*, specific biochemical and genetic tests are needed to distinguish *Shigella*. Techniques for detecting and estimating numbers of *Shigella* in a food sample have been described by Andrews and Jacobson (2013) and Lampel and Zhang (2015). The sensitivity of testing methods is important to be able to detect low levels of *Shigella* in foods.

Controls: Effective controls to lower the risk of foodborne illness from this species include the following:

- Good agricultural, sanitation, and hygiene practices will limit the number of bacteria present in a food.
- Temperature control that keeps food out of the temperature danger zone (4–60°C; 40–140°F) is important to prevent growth.
- Acidic foods with pH values lower than 4.6 are low risk for illness due to this pathogen.
- *Shigella* are sensitive to high salt concentrations, ionizing radiation, and cooking.

Vibrio

Microscopic morphology: This genus of Gram-negative bacteria is rod shaped and can be straight or curved. The curved shape resembles a comma (see Fig. 7.4). They are motile bacteria with a long flagellum. The genus name comes from Filippo Pacini, who in 1854 described vibrioni in his microscopic examination of specimens from sufferers of cholera (Howard-Jones, 1984).

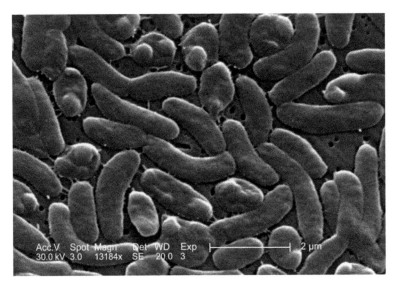

Fig. 7.4. Scanning electron micrograph showing a grouping of *Vibrio vulnificus* (CDC PHIL, 2014; Janice Carr).

Environmental factors: *Vibrio* are aquatic bacteria. Their natural environment includes coastal waters and estuaries. Not all are pathogenic. The highest risk foods for consuming infectious *Vibrio* are shellfish that are filter feeders (see Box 7.5). These animals tend to concentrate *Vibrio* bacteria from the infested seawater into their tissues, and are of particular high risk for foodborne illness when consumed raw.

Vibrio are considered mesophiles. In their natural environment of seawater, they multiply when the temperature is between 19 and 20°C (68°F). *Vibrio parahemolyticus*, which is the most common species associated with cross-contamination of food, has a growth range of 5–44°C (41–111.2°F), with an optimum growth range of 30–35°C (86–95°F).

Vibrio are mildly, but obligatory, halophilic and require sodium chloride concentrations of 1 to 8% to multiply. They require a minimum water activity of 0.93. Their generation time under optimal conditions can be quite rapid, 9 to 13 minutes. They can be challenging to culture and are often in a viable but non-culturable state.

Type of foodborne illness: There are over 80 species within the genus, and nine of these are associated with foodborne illness outbreaks (Oliver and Kaper, 2007). These nine species are *V. cholerae, V. parahaemolyticus, V. vulnificus, V. mimicus* (similar to *cholerae*), *V. alginolyticus, V. fluvialis, V. furnissii, V. metschnikovii,* and *V. holisae*.

V. cholerae is the species responsible for human cholera. Cholera is a toxin-mediated infection, and these infectious bacteria are spread rapidly through polluted water. There have been seven pandemics of cholera in recorded history. A pandemic is a global disease outbreak. Simply put, it is when an outbreak crosses continental borders. The World Health Organization monitors pandemics and offers emergency response guidance (WHO, 2016). These seven pandemics all originating from Asia, and the world has been in the seventh pandemic since 1961, with periodic large outbreaks (Kaper *et al.*, 1995).

Cholera is endemic in approximately 50 countries and can spread quickly when outbreaks emerge (Routh *et al.*, 2015). Cholera is a common concern after a natural disaster such as an earthquake or typhoon that compromises clean water supplies. Cholera is spread through many mechanisms, including foodborne. An epidemic in Central and South America that started in Peru in 1991 was thought to be caused from ballast water from ships from the orient.

Cholera manifests in the small intestine where cholera toxins excreted by *V. cholerae* reverse the flow of electrolytes and water absorption to secretion. This results in very voluminous, watery diarrhea. The volume of diarrhea can exceed 20 liters/day and death can come rapidly if untreated. For untreated cases, mortality rates range from 50 to 60%.

Box 7.5. *Vibrio* – what's in your oysters?

There is an old myth that oysters should be eaten in months that contain an R, like September through April. That leaves May through August to avoid oysters. This seems somewhat logical since we might expect to see higher concentrations of bacteria like *Vibrio* in waters where oysters are harvested from during warmer months. However, *Vibrio* can be present in any month, and immunocompromised people should be especially wary of eating oysters raw (SafeOysters.org, 2009).

All coastal waters and estuaries contain *Vibrio* bacteria. Although many *Vibrio* species are harmless, several can cause serious disease in humans or animals. *Vibrio vulnificus* and *V. parahaemolyticus* are the most common types of sometimes deadly foodborne and wound *Vibrio* infections. Recognized infections from *Vibrio* species are on the rise, and although there is some uncertainty, most researchers predict that climate change will increase cases (Froelich and Noble, 2016).

According to these authors, *Vibrio* population concentrations do correlate with seasonal temperatures, but regional environmental and strain variations complicate predictions of risk, making it challenging to effectively manage oyster harvesting practices. There are many non-pathogenic *Vibrio* in these ecosystems, so the risk control measures need to adjust for the actual presence and concentrations of pathogenic *Vibrio* strains. Control measures typically involve harvest restrictions in identified areas for periods of time until acceptable concentrations are detected. Consumer education is another risk control approach but has limited effectiveness. Raw oysters are considered a delicacy by many consumers, and convincing people to consider the risks can be challenging. The risk level for immunocompromised individuals to become gravely ill from consuming raw oysters is high. Regulations and education campaigns help to lower risk, but the rate of infections in the United States continues to increase.

V. parahaemolyticus causes an infection that leads to gastroenteritis. The onset of symptoms is 3 to 76 hours, with a duration range of 1 to 8 days. Symptoms include diarrhea (sometimes bloody), cramps, weakness, nausea, chills, headache, and vomiting. The minimum infectious dose is about 10^5 cells. Incidence is low in the United States and Europe. Incidence is highest in Japan where it accounts for 70% of the bacterial foodborne illnesses.

V. vulnificus is a highly invasive pathogen that causes a toxin-mediated infection. The bacteria produce toxins that allow tissue invasion leading to soft-tissue infection and septicemia. This species accounts for 95% of seafood-related deaths in the United States, and of those, immunocompromised people are most affected, particularly those with liver diseases. The highest incidence rates are in coastal areas.

Foodborne illness statistics: *Vibrio* are estimated to cause 52,408 illnesses, 278 hospitalizations, and 48 deaths per year in the United States (Scallon *et al.*, 2011). Of these, *V. cholerae* is estimated to cause 84 illnesses, two hospitalizations, and zero deaths per year; *V. vulnificus* is estimated to cause 96 illnesses, 93 hospitalizations, and 36 deaths per year; *V. parahaemoliticus* is estimated to cause 34,664 illnesses, 100 hospitalizations, and four deaths per year; and all other *Vibrio* spp. are estimated to cause 17,564 illnesses, 83 hospitalizations, and eight deaths per year. *V. vulnificus* is the most fatal foodborne pathogen in the United States.

Costs of foodborne illness: The cost per case for *Vibrio* was estimated at US$2,226/case for *V. cholerae*, US$2,792,171/case for *V. vulnificus*, US$2,551/case for *V. parahaemolyticus*, and US$5,020/case for *all other Vibrio* spp. (Scharff, 2012).

Testing methods: Techniques for detecting and estimating numbers of *Vibrio* in a food sample have been described by Kaysner and DePaola (2004) and DePaola and Jones (2015). Sensitivity of testing methods is important to be able to detect low levels of *Vibrio* in foods. *Vibrio* are difficult to culture. They either lose viability in samples or are commonly in a viable but unculturable state. The application of DNA-based methods of detection of *Vibrio* has several advantages over traditional culture methods.

Controls: Effective controls to lower the risk of food-borne illness from this species include the following:

- Good sanitation and hygiene practices will limit cross-contamination.
- Temperature control that keeps food out of the temperature danger zone (4–60°C; 40–140°F) is important to prevent growth.
- *Vibrio* are sensitive to ionizing radiation and cooking.
- Monitoring of environmental loads of *Vibrio* can help with management of seafood harvesting to prevent contamination.

Yersinia

Microscopic morphology: *Yersinia* spp. are Gram-negative bacteria with a straight or curved rod shape, similar to *Vibrio*. The genus name is derived from the discoverer, French bacteriologist Alexandre Emil John Yersin (Haubrich, 2004).

Environmental factors: *Yersinia* can be found widely distributed in the environment, and are commonly isolated from soil, lakes, wells, and streams. They can be found in the digestive tracts of warm-blooded animals. Swine are the most common source for *Yersinia* that are pathogenic to humans. *Yersinia* have also been spread through raw milk.

Yersinia are categorized as psychrotrophs. They have a wide temperature growth range of –2 to 45°C (28.4–113°F), and an optimum growth temperature range of 22–29°C (71.6–84.2°F).

Yersinia can multiply in a pH range of 4.6 to 9.0, with an optimum pH range of 7 to 8. They can also tolerate sodium chloride concentrations as high as 5%.

Type of foodborne illness: *Yersinia* cause infections. There are 11 species of *Yersinia*, of which three are pathogenic to humans: *Y. enterocolitica* (foodborne illness), *Y. pseudotuberculosis* (foodborne illness), and *Y. pestis* (cause of the plague; not foodborne).

Foodborne infections are initiated with binding to the small intestine mucosal layer. The bacteria then enter the lymphatic tissue and can spread through infected monocytes. Infectious doses are low in virulent strains, and are estimated at about 100 cells. As with *Campylobacter, Salmonella,* and *Shigella* infections, a post-infectious reactive arthritis can occur in some cases of yersiniosis. Other post-infection complications can manifest as chronic ailments including:

- erythema nodosum (tender nodules usually on the legs);
- iridocyclitis (inflammation of the eye);
- glomerulonephritis;
- carditis; and
- thyroiditis.

The highest incidence rates seem to occur in autumn. Symptoms of foodborne yersiniosis include fever, headache, diarrhea, vomiting, severe abdominal pain (can be confused with appendicitis), and pharyngitis (sore throat). The onset time for symptoms ranges from 24 to 36 hours, and can take even longer. The duration of symptoms ranges from 1 to 3 days.

Box 7.6. *Yersinia* – tales of cross-contamination

Yersiniosis outbreaks are fairly rare in the United States. For an example of a fairly large outbreak, let's go back to 1982 (CDC, 1982). Between the months of June and July, a large outbreak of enteritis caused by *Y. enterocolitica* occurred across the states of Arkansas, Tennessee, and Mississippi. This included 172 identified cases, of which 41% were children less than 5 years of age. Most of the victims were hospitalized and 17 underwent appendectomies. There were likely more unidentified cases involved.

The strain of *Y. enterocolitica* involved was linked to a dairy processing plant in Tennessee. Pasteurized milk was implicated. The exact source was never publicly identified, and the outbreak ended spontane-

ously. It could be that the bacteria were present in large enough numbers in the raw milk for some to have survived pasteurization followed by growth at refrigeration temperatures. A more likely scenario would be from cross-contamination post-processing.

An outbreak of yersiniosis from 1976 also demonstrates how cross-contamination can lead to many serious cases of illness (Black *et al.*, 1978). This outbreak occurred from October through September of 1976. The 38 cases were all school children, and 16 of them had appendectomies. The source was traced to chocolate milk served in school cafeterias where the contamination was introduced post-pasteurization through hand mixing of the chocolate into the milk.

Considering the environmental presence of *Yersinia*, outbreaks are uncommon (see Box 7.6). The majority of outbreaks tend to occur in Europe, particularly in Scandinavia. They also seem to cluster in Canada.

Foodborne illness statistics: *Y. enterocolitica* is estimated to cause 97,656 illnesses, 533 hospitalizations, and 29 deaths per year in the United States (Scallon *et al.*, 2011).

Costs of foodborne illness: Overall, the estimated costs was US$11,334/case in the United States in 2010 (Scharff, 2012).

Testing methods: Techniques for detecting and estimating numbers of *Yersinia* in a food sample have been described by Weagant and Feng (2007) and Ceylan (2015). The sensitivity of testing methods is important to be able to detect low levels of *Yersinia* in foods. Because of their psychrotrophic nature, cold enrichment can be used in isolation methods.

Controls: Effective controls to lower the risk of foodborne illness from this species include the following:

- Good agricultural, sanitation, and hygiene practices will limit the number of bacteria present in a food.
- Temperature control that keeps food out of the temperature danger zone (4–60°C; 40–140°F) is important to prevent and slow growth.
- *Yersinia* are sensitive to ionizing radiation and cooking.

Summary

The difference between classification of bacteria as Gram positive or Gram negative was explained in Chapter 6. In this chapter, six Gram-negative bacteria of most concern for foodborne illness were described: *Campylobacter*, *E. coli*, *Salmonella*, *Shigella*, *Vibrio*, and *Yersinia*. Pathogenic strains within each of these six genera are commonly spread through fecal contamination, making personal hygiene of food handlers an important control point, along with prevention of cross-contamination.

Further Reading

FDA (2012) *Bad Bug Book*, 2nd edn. United States Food and Drug Administration, Washington, DC, USA. Available at: http://www. fda.gov/Food/FoodborneIllnessContaminants/ CausesOfIllnessBadBugBook/default.htm (accessed 29 December 2015).

Juneja, V.K. and Sofos, J.N. (2010) *Pathogens and Toxins in Foods: Challenges and Interventions.* ASM Press, Washington, DC, USA.

References

Ajene, A.N., Fischer Walker, C.L., and Black, R.E. (2013) Enteric pathogens and reactive arthritis: a systematic review of *Campylobacter*, *Salmonella* and *Shigella*-associated reactive arthritis. *Journal of Health, Population, and Nutrition* 3, 299–307.

Andrews, W.H. and Jacobson, A. (2013) Chapter 6. *Shigella*. In: *Bacteriological Analytical Manual (BAM).* United States Food and Drug Administration, Washington, DC, USA. Available at: http://www.fda.gov/ Food/FoodScienceResearch/LaboratoryMethods/ ucm2006949.htm (accessed 29 December 2015).

Andrews, W.H, Jacobson, A., and Hammack, T. (2015) Chapter 5. *Salmonella*. In: *Bacteriological Analytical Manual (BAM).* United States Food and Drug Administration, Washington, DC, USA. Available at: http://www.fda.gov/Food/FoodScienceResearch/ LaboratoryMethods/ucm2006949.htm (accessed 29 December 2015).

Behravesh , C.B., Mody, R.K., Jungk, J., Gaul, L., Redd, J.T., Chen, S., Cosgrove, S., Hedican, E., Sweat, D., Chavez-Hauser, L., Snow, S.L. Hanson, H., Nguyen, T.A., Sodha, S.V., Boore, A.L., Russo, E., Mikoleit, M., Theobald, L., Gerner-Smidt, P., Hoekstra, R.M., Angulo, F.J., Swerdlow, D.L., Tauxe, R.V., Griffin, P.M., and Williams, I.T. (2011) 2008 outbreak of *Salmonella* Saintpaul infection associated with raw produce. *New England Journal of Medicine* 364, 918–27.

Black, R.E., Jackson, R.J., Tsai, T., Madvesky, M., Shayegani, M., Freeley, J.C., MacLeod, K.I.E., and Wakelee, A.M. (1978) Epidemic *Yersinia enterocolitica* infection due to contaminated chocolate milk. *New England Journal of Medicine* 298, 76–79.

Bowen, A. (2015) Chapter 3. Infectious diseases related to travel: shigellosis. In: *The Yellow Book: CDC Health Information for International Travel 2016*. United States Centers for Disease Control and Prevention, Washington, DC, USA. Available at: http://wwwnc.cdc. gov/travel/page/yellowbook-home-2014 (accessed 29 December 2015).

Brenner, F.W., Villar, R.G., Angulo, R.T., and Swaminathan, B. (2000) *Salmonella* nomenclature. *Journal of Clinical Microbiology* 38, 2465–2467.

CDC (1982) Epidemiologic notes and reports multi-state outbreak of yersionsis. *Morbidity and Mortality Weekly Report* 31, 505–506.

CDC (1998) Multistate outbreak of *Salmonella* serotype Agona infections linked to toasted oats cereal –

United States, April–May, 1998. *Morbidity and Mortality Weekly* 47, 462–464.

CDC (2002) Multistate outbreak of *Escherichia coli* O157:H7 infections associated with eating ground beef – United States, June–July 2002. *Morbidity and Mortality Weekly Report* 51, 637–639.

CDC (2008) Outbreak of *Salmonella* serotype Saintpaul infections associated with multiple raw produce items – United States, 2008. *Morbidity and Mortality Weekly Report*. 57, 929–934.

CDC (2009) Multistate outbreak of *Salmonella* Typhimurium infections linked to peanut butter, 2008–2009. Available from http://www.cdc.gov/salmonella/2009/peanut-butter-2008-2009.html (accessed 29 December 2015).

CDC (2013) Outbreak of *Escherichia coli* O104:H4 infections associated with sprout consumption – Europe and North America, May–July 2011. *Morbidity and Mortality Weekly Report* 62, 1029–1031.

CDC (2015a) *E. coli* (*Escherichia coli*): general information. Available at: http://www.cdc.gov/ecoli/general/index.html/ (accessed 29 December 2015).

CDC (2015b) What is salmonellosis? Available at: http://www.cdc.gov/Salmonella/general/index.html (accessed 29 December 2015).

CDC PHIL (2014) United States Centers for Disease Control Public Health Image Library. Available at: http://phil.cdc.gov/phil/home.asp (accessed 29 December 2015).

Ceylan, E. (2015) Chapter 41. *Yersinia*. In: Salfinger, Y. and Tortorello, M.L. (eds) *Compendium of Methods for the Microbiological Examination of Foods.* Association of Public Health Laboratories, Washington, DC, USA, pp. 445–475.

Chen, H.M., Wang, Y., Su, L.H., and Chiu, C.H. (2013) Nontyphoid *Salmonella* infection: microbiology, clinical features, and antimicrobial therapy. *Pediatrics and Neonatology* 54, 147–152.

Connor, B.A. (2015) Chapter 2. The pre-travel consultation. In: *The Yellow Book: CDC Health Information for International Travel 2016.* United States Centers for Disease Control and Prevention, Washington, DC, USA. Available at: http://wwwnc.cdc.gov/travel/page/yellowbook-home-2014 (accessed 29 December 2015).

Corry, J.E.L. and Line, J.E. (2015) Chapter 30. *Campylobacter*. In: Salfinger, Y. and Tortorello, M.L. (eds) *Compendium of Methods for the Microbiological Examination of Foods.* Association of Public Health Laboratories, Washington, DC, USA, pp. 365–373.

Cox, N.A., Frye, J.G., McMahon, W., Jackson, C.R., Richardson, J., Cosby, D.E., Mead, G., and Doyle, M.P. (2015) Chapter 36. *Salmonella*. In: Salfinger, Y. and Tortorello, M.L. (eds) *Compendium of Methods for the Microbiological Examination of Foods.* Association of Public Health Laboratories, Washington, DC, USA, pp. 445–475.

Croxen, M.A. and Finlay, B.B. (2010) Molecular mechanisms of *Escherichia coli* pathogenicity. *Nature Reviews Microbiology* 8, 26–38.

Delannoy, S., Beutin, L. and Patrick, F. (2013) Discrimination of enterohemorrhagic *Escherichia coli* (EHEC) from non-EHEC strains based on detection of various combinations of type III effector genes. *Journal of Clinical Microbiology* 51, 3257–3262.

DePaola, A. and Jones, J.L. (2015) Chapter 40. *Vibrio*. In: Salfinger, Y. and Tortorello, M.L. (eds) *Compendium of Methods for the Microbiological Examination of Foods.* Association of Public Health Laboratories, Washington, DC, USA, pp. 103–120.

Eiko, Y. (2002) *Bacillus dysentericus* (sic) 1897 was the first taxonomic rather *than Bacillus dysenteriae* 1898. *International Journal of Systematic and Evolutionary Microbiology* 52, 1041.

FDA (2016) Questions and answers on improving the safety of spices. Available at: http://www.fda.gov/Food/FoodScienceResearch/RiskSafetyAssessment/ucm487954.htm (accessed 29 February 2016).

Feng, P., Weagant, S.D., Grant, M.A., and Burkhardt, W. (2013) Chapter 4. Enumeration of *Escherichia coli* and the coliform bacteria. In: *Bacteriological Analytical Manual (BAM).* United States Food and Drug Administration, Washington, DC, USA. Available at: http://www.fda.gov/Food/FoodScienceResearch/LaboratoryMethods/ucm2006949.htm (accessed 29 December 2015).

Feng, P., Weagant, S.D., and Jinneman, K. (2015) Chapter 4A. Diarrheagenic *Escherichia coli*. In: *Bacteriological Analytical Manual (BAM).* United States Food and Drug Administration, Washington, DC, USA. Available at: http://www.fda.gov/Food/FoodScienceResearch/LaboratoryMethods/ucm2006949.htm (accessed 29 December 2015).

Froelich, B.A. and Noble, R.T. (2016) *Vibrio* bacteria in raw oysters: managing risks to human health. *Philosophical Transactions of the Royal Society B* 371, 20150209.

FSIS (2015) Safe minimum internal temperature chart. United States Department of Agriculture, Food Safety and Inspection Service. Available at: http://www.fsis.usda.gov/wps/portal/fsis/topics/food-safety-education/get-answers/food-safety-fact-sheets/safe-food-handling/safe-minimum-internal-temperature-chart/ct_index (accessed 30 December 2015).

Gal-Mor, O., Boyle, E.C., and Grassel, G.A. (2014) Same species, different diseases: how and why typhoidal and non-typhoidal *Salmonella enterica* serovars differ. *Frontiers in Microbiology* 5, 391.

Haubrich, W.S. (2004) Yersin of *Yersinia* infection. *Gastroenterology* 128, 23.

Hennessy, T.W., Hedberg, C.W., Slutsker, L., White, K.E., Besser-Wiek, J.M., Moen, M.E., Feldman, J., Coleman, W.W., Edmonson, L.M., MacDonald, K.L., and Osterholm, M.T. (1996) A national outbreak of *Salmonella enteritidis* infections from ice cream. The Investigation Team. *New England Journal of Medicine* 334, 1281–1286.

Howard-Jones, N. (1984) Robert Koch and the cholera vibrio: a centenary. *British Medical Journal* 288, 379–381.

Hunt, J.M., Abeyta, C., and Tran, T. (2001) Chapter 7. *Campylobacter*. In: *Bacteriological Analytical Manual (BAM)*. United States Food and Drug Administration, Washington, DC, USA. Available at: http://www.fda.gov/Food/FoodScienceResearch/LaboratoryMethods/ucm2006949.htm (accessed 29 December 2015).

Iwamoto, M. (2015) Chapter 3. Infectious diseases related to travel: salmonelosis. In: *The Yellow Book: CDC Health Information for International Travel 2016*. United States Centers for Disease Control and Prevention, Washington, DC, USA. Available at: http://wwwnc.cdc.gov/travel/page/yellowbook-home-2014 (accessed 29 December 2015).

Janet O. (2015) Owner of SJ Restaurant linked to *Shigella* outbreak speaks out. Available at: http://abc7news.com/health/owner-of-sj-restaurant-linked-to-shigella-outbreak-speaks-out/1072017/ (accessed 29 December 2015).

Kaper, J.B., Morris, J.G., Levine, M.M. (1995) Cholera. *Clinical Microbiology Reviews* 8, 48–86.

Kaysner, C.A. and DePaola, A. (2004) Chapter 9. *Vibrio*. In: *Bacteriological Analytical Manual (BAM)*. United States Food and Drug Administration. Washington, DC, USA. Available at: http://www.fda.gov/Food/FoodScienceResearch/LaboratoryMethods/ucm2006949.htm (accessed 29 December 2015).

Kornacki, J.L., Gurtler, J.B., and Stawick, B.A. (2015) Chapter 9. Enterobacteriaceae, coliforms, and *Escherichia coli* as quality and safety indicators. In: Salfinger, Y. and Tortorello, M.L. (eds) *Compendium of Methods for the Microbiological Examination of Foods*. Association of Public Health Laboratories, Washington, DC, USA, pp. 103–120.

Lampel, K.A. and Zhang, G. (2015) Chapter 37. *Shigella*. In: Salfinger, Y. and Tortorello, M.L. (eds) *Compendium of Methods for the Microbiological Examination of Foods*. Association of Public Health Laboratories, Washington, DC, USA, pp. 103–120.

Malbury, T. (2015) FSM scoop: raw sprouts. *Food Safety Magazine*. Available at: http://www.foodsafetymagazine.com/enewsletter/fsm-scoop-raw-sprouts/ (accessed 29 December 2015).

Meng, J., Fratamico, P.M., and Feng, P. (2015) Chapter 34. Pathogenic *Escherichia coli*. In: Salfinger, Y. and Tortorello, M.L. (eds) *Compendium of Methods for the Microbiological Examination of Foods*. Association of Public Health Laboratories, Washington, DC, USA, pp. 411–424.

Newton, A.E., Routh, J.A., and Mahon, B.E. (2015) Chapter 3 Infectious diseases related to travel: typhoid & paratyphoid fever. In: *The Yellow Book: CDC Health Information*

for International Travel 2016*. United States Centers for Disease Control and Prevention, Washington, DC, USA. Available at: http://wwwnc.cdc.gov/travel/page/yellowbook-home-2014 (accessed 29 December 2015).

Nygren, B.L., Schilling, K.A., Glanton, M.A., Silk, B.J., Cole, D.J., and Mintz, E.D. (2015) Foodborne outbreaks of shigellosis in the USA, 1998–2008. *Epidemiology and Infection* 141, 233–241.

Oliver, J.D. and Kaper, J.B. (2007) Chapter 16. *Vibrio* species. In: Doyle, M.P. and Beuchat, L.R. (eds) *Food Microbiology: Fundamental and Frontiers,* 3rd edn. ASM Press, Washington, DC, USA, pp. 343–379.

Rasko, D.A., Webster, D.R., Sahl, J.W., Bashir, A., Boisen, N., Scheutz, F., Paxinos, E.E., Sebra, R., Chin, C.S., Iliopoulos, D., Klammer, A., Peluso, P., Lee, L., Kislyuk, A.O., Bullard, J., Kasarskis, A., Wang, S., Eid, J., Rank, D., Redman, J.C., Steyert, S.R., Frimodt-Møller, J., Struve, C., Petersen, A.M., Krogfelt, K.A., Nataro, J.P., Schadt, E.E., and Waldor, M.K. (2011) Origins of the *E. coli* strain causing an outbreak of hemolytic–uremic syndrome in Germany. *New England Journal of Medicine* 365, 709–717

Reller, M.E., Nelson, J.M., Molbak, K., Ackman, D.M., Schoonmaker-Bopp, D.J., Root, T.P., and Mintz, E.D. (2006) A large, multiple-restaurant outbreak of infection with *Shigella flexneri* serotype 2a traced to tomatoes. *Clinical Infectious Disease* 42, 163–169.

Routh, J.A., Newton, A.E. and Mintz, N.E. (2015) Chapter 3. Infectious diseases related to travel: cholera. In: *The Yellow Book: CDC Health Information for International Travel 2016*. United States Centers for Disease Control and Prevention, Washington, DC, USA. Available at: http://wwwnc.cdc.gov/travel/page/yellowbook-home-2014 (accessed 29 December 2015).

Salmon, D.E. and Smith, T (1885) Report on swine plague. In: *USDA Bureau of Animal Industry 2nd Annual Report*. United States Department of Agriculture, Washington, DC, USA.

Scallon, E., Hoekstra, R.M., Angulo, F.J., Tauxe, R.V., Widdowson, M.A., Roy, S.L., Jones, J.L., and Griffin, P.M. (2011) Foodborne illness acquired in the United States – major pathogens. *Emerging Infectious Diseases* 17, 7–15.

Scharff, R.L. (2012) Economic burden from health losses due to foodborne illness in the United States. *Journal of Food Protection* 75 123–131.

Taylor, E.V., Herman, K.M., Ailes, E.C., Fitzgerald, C., Yoder, J.S., Mahon, B.E., and Tauxe, R.V. (2012) Common source outbreaks of *Campylobacter* infection in the USA, 1997–2008. *Epidemiology and Infection* 141, 987–996.

USDA (2013) Color of cooked ground beef as it relates to doneness. Available at: http://www.fsis.usda.gov/wps/portal/fsis/topics/food-safety-education/get-answers/food-safety-fact-sheets/meat-preparation/color-of-cooked-ground-beef-as-it-relates-to-doneness/CT_Index (accessed 29 December 2015).

Weagant, S.D. and Feng, P. (2007) Chapter 8. *Yersinia enterocolitica*. In: *Bacteriological Analytical Manual (BAM)*. United States Food and Drug Administration, Washington, DC, USA. Available at: http://www.fda.gov/Food/FoodScienceResearch/LaboratoryMethods/ucm2006949.htm (accessed 29 December 2015).

WHO (2016) Emergencies preparedness, response. Available at: http://www.who.int/csr/disease/en/ (accessed 1 February 2016).

Yuki, N. and Hartung, H.P. (2012) Medical progress: Guillain–Barré syndrome. *New England Journal of Medicine* 366, 2294–2304.

Zuraw, L. (2015) The 10 biggest U.S. foodborne illness outbreaks of 2015. Available at http://www.foodsafetynews.com/2015/12/the-10-biggest-u-s-foodborne-illness-outbreaks-of-2015/#.VthNIU32ZaR (accessed 29 December 2015).

8 Eukaryotic Microorganisms of Concern in Food: Parasites and Molds

> **Key Questions**
> - What is a eukaryotic microorganism?
> - What eukaryotic microorganisms are of most concern for microbial food safety?
> - What are the mechanisms by which these microorganisms cause illness?
> - What are the hazards these microorganisms present to consumers?
> - What controls can prevent foodborne illness due to these microorganisms?

Eukaryotic Microorganisms of Concern in Foods

In biology, eukaryote is a term that describes organisms that are a cell or have cells with nuclei contained within a nuclear membrane, and other membrane-bound cellular organelles. These are distinct from prokaryotes, which are single-celled microorganisms that do not have a nucleus or any other membrane-bound cellular organelles, like the bacteria discussed in Chapter 6 and 7. The eukaryotes of concern in food safety are tiny microscopic creatures.

The eukaryotic microorganisms of most concern in foods discussed in this chapter include parasites from the Animalia and Protista kingdoms and molds from the kingdom of Fungi. The parasites cause parasitic infections, and the molds cause intoxications via preformed mycotoxins in the food.

The plant kingdom also includes eukaryotic microscopic algae that produce toxins that can bioaccumulate in aquatic food chains, affecting seafood safety. Since food animals are the transmission source for these toxins, they are not included here. For more information about toxin-producing algae that impact seafood safety, see NOAA (2007).

Foodborne Parasites: Helminths and Protozoa

A parasite is an organism that must live off a living host organism. The living host is required for food and often for the proper reproductive environment for the parasite. A successful parasite does not overtax the host. Parasites that harm or kill the host can be successful as well, if the result facilitates transmission to a new host.

Parasites that can be transmitted through food include some helminths and protozoa. Their transmissible forms range from tiny to macroscopic in size.

> Protozoa are microscopic, one-celled organisms that can be free-living or parasitic in nature. They are able to multiply in humans, which contributes to their survival and also permits serious infections to develop from just a single organism. Transmission of protozoa that live in a human's intestine to another human typically occurs through a fecal-oral route (for example, contaminated food or water or person-to-person contact) ... Helminths are large, multicellular organisms that are generally visible to the naked eye in their adult stages. Like protozoa, helminths can be either free-living or parasitic in nature (CDC, 2014).

These microorganisms only need to survive in food as a transport mechanism between hosts. They do not multiply in food; they just need to survive the intrinsic and extrinsic factors of the food long enough to enter the human host. Parasitic infections may account for about 2–7% of foodborne infections in the United States (Orlandi *et al.*, 2002; Scallon *et al.*, 2011). "Globalization of food trade, preferences for raw and undercooked dishes, ease of international travel, and increasing numbers of immunocompromised individuals are factors that have contributed to the increase in foodborne parasitic infections" (Orlandi *et al.*, 2002).

Helminth foodborne parasites

This group of parasites includes flatworms and roundworms. Many soil-transmitted helminths can be transmitted through food. The World Health Organization estimates that more than 1 billion people worldwide are infected with one or more of these parasites (WHO, 2016). Roundworms range from less than 1 mm to 7 m in length and have a cylindrical body with a mouth and anus at opposite ends, while flatworms can range in length from 1 mm to 20 m and have a flattened shape and one opening that serves as both a mouth and anus.

Anisakis simplex

Microscopic morphology: This parasite is an example of a roundworm. The adult worm form averages 2–3 cm in length (but can reach 5–20.5 cm in length), and is the diameter of a human hair. The genus name *Anisakis* is derived from Greek words that describe unequal pointed objects. This is descriptive of the mouth portion of this roundworm (Fig. 8.1).

Environmental factors: *A. simplex* is a parasite of marine mammals, moving through its various life-cycle forms through planktonic crustaceans, to larger fish and shellfish, and then finally to a sea mammal as the final host (Audicana and Kennedy, 2008). The parasite lays eggs while attached in the final host, and the host excretes these eggs back into the marine environment with its feces. Humans are not a normal definitive host, but can become infected through consumption of infected marine animals. Also known as the cod worm or herring worm, *A. simplex* is found in raw or undercooked seafood. Common methods of preparing raw seafood include salting, curing, marinating, pickling, and cold smoking, but these methods are not sufficient to kill this parasite.

Type of foodborne illness: The parasitic infection caused by *A. simplex* initiates with swallowing at least one live worm. One live worm can be all it takes for an infectious dose. Within 1 to 12 hours of consuming a live worm, symptoms of acute abdominal pain, nausea, vomiting, and diarrhea can occur. Coughing can be a symptom if the worm attaches high up in the throat upon swallowing. Some infections are asymptomatic, while others can result in severe inflammation of the digestive tract. In some untreated infections, ulcer-like symptoms can occur. In rare cases, complications such as small bowel obstruction and intestinal perforation can occur (Hochberg and Hamer, 2010). Some human hosts experience allergic responses in conjunction with the infection (Audicana and Kennedy, 2008).

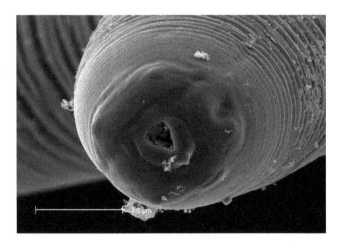

Fig. 8.1. A scanning electron micrograph of the mouthparts of *Anisakis simplex* (adapted from https://en.wikipedia.org/wiki/File:So4b-08.jpg; public domain).

Foodborne illness statistics: Infections by *A. simplex* are rare in the United States, and statistical information is limited. These parasitic infections are much more common in Japan and northern Europe where a larger proportion of the diet includes fresh seafood. About 90% of reported cases have happened in Japan (Fayer and Xiao, 2015). Hochberg and Hamer (2010) attribute the rising global incidence to larger host populations, increased consumption of raw and undercooked seafood, and improved endoscopic diagnostic techniques.

Costs of foodborne illness: There is a lack of information in the literature regarding the cost per case for this type of foodborne parasitic infection.

Testing methods: Bier *et al.* (2001) have described the methods for detecting the parasite larvae and eggs in foods. Fayer and Xiao (2015) have described the methods used for detecting parasites, including anisakids, in mollusks, crustaceans, fish, and frog legs. Diagnostic images are available from the United States Centers for Disease Control (CDC, 2013). Traditional methods are typically low-tech and involve candling or dissolving of the muscle tissue to reveal the parasite.

Controls: Effective controls to lower the risk of foodborne illness from this species include the following:

- The parasite is sensitive to cooking, freezing, and irradiation. However, dead parasites may still be able to induce allergic responses (FDA, 2001).

- Seafood should be thoroughly inspected for the presence of parasites.
- Avoiding cross-contamination is important, especially between raw meat and ready-to-eat foods. Education of food handlers, including consumers, is important to ensure an understanding of how to prevent cross-contamination.

Taenia

Microscopic morphology: *Taenia* spp. are flatworms. The genus name is derived from the Greek word *tainia*, meaning ribbon. Species differentiation is based on the morphology of the so-called head and mouth portions of the worms (CDC, 2013). These flatworms range in size from about 5 mm to 25 m (Fig. 8.2).

Environmental factors: There are three major types of *Taenia* that infect humans: *Taenia solium* (the pork tapeworm); *Taenia asiatica* (the Asian tapeworm, also the pork tapeworm); and *Taenia saginata* (the beef tapeworm).

These species are very host adapted and need the animal and human species to host their various life cycle forms. Consumption of raw or undercooked beef or pork results in transmission to the human host. Rates are highest in areas with poor sanitation and in places that use human waste (night soil) for fertilizer.

Type of foodborne illness: One larva is enough for an infectious dose to cause taeniasis. The onset

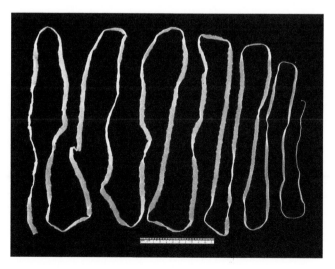

Fig. 8.2. An adult *Taenia saginata* tapeworm (CDC PHIL, 2014). The ruler at the bottom of the photograph is 11.5 cm long in cm increments.

time to symptoms can range from 8 to 14 weeks. It takes about 2 months for the worm to mature in the human small intestine. Symptoms include abdominal discomfort, weight loss, anorexia, nausea, insomnia, weakness, perianal pruritus, and nervousness. Mortality is rare, and the specific rates are unknown.

A more severe form of parasitic infection can occur with the *T. solium* species. The larval cysts can invade human tissue, including the brain, and cause cysticercosis. In severe cases of cysticercosis, death can occur.

Foodborne illness statistics: Most cases occur in Latin America, Africa, and South/South-east Asia, with lower rates of occurrence in eastern Europe, Spain, and Portugal (Cantey and Jones, 2015). Documented infections are rare in the United States, and statistical information is limited.

> Higher rates of illness have been seen in people in Latin America, Eastern Europe, sub-Saharan Africa, India, and Asia. *Taenia solium* taeniasis is seen in the United States, typically among Latin American immigrants. *Taenia asiatica* is limited to Asia and is seen mostly in the Republic of Korea, China, Taiwan, Indonesia, and Thailand (CDC, 2014).

Costs of foodborne illness: There is a lack of information in the literature regarding the cost per case for this type of foodborne parasitic infection.

Testing methods: Bier *et al.* (2001) have described the methods for detecting parasite larvae and eggs in foods. Fayer and Xiao (2015) have described the methods used for detecting parasites, and Ito and Craig (2003) have described immunogenic and molecular methods. Diagnostic images are available from the United States Centers for Disease Control (CDC, 2013).

Controls: Effective controls to lower the risk of foodborne illness from this species include the following:

- The parasite is sensitive to cooking, freezing, and irradiation. Cooking various types of meat to appropriate internal temperatures, i.e. 160°C (320°F) for ground meat, is recommended (CDC, 2014).

- Avoiding cross-contamination is important, especially between raw meat and ready-to-eat foods. Education of food handlers, including consumers, is important to ensure an understanding of how to prevent cross-contamination.

Trichinella spiralis

Microscopic morphology: *T. spiralis* is a roundworm. The genus name comes from the Greek word *trikhine*, which means like hair. The species name is descriptive of how this roundworm curls up on itself in a spiral. The length of an adult male worm varies from 1.2 to 1.6 mm, and is double for female worms.

Environmental factors: Almost any warm-blooded carnivore or omnivore can serve as a host for this parasite. Transmission is through consumption of raw or undercooked muscle tissue that contains encysted larvae. The larvae, upon digestion of the cyst in the small intestine of the host, are released and deposited in the mucosal layer. They develop into the adult stage and reproduce (Fig. 8.3). The females release larvae that burrow into the muscle tissue to encyst (CDC, 2013).

This parasite is found worldwide, but is most commonly found in regions of Europe and the United States. For infections in humans, the most common types of contaminated meats consumed to transmit the parasite include pork, horse, bear (see Box 8.1), and other wild game meat (Fayer and Xiao, 2015).

Type of foodborne illness: Mild symptoms can result from the intestinal colonization, and include diarrhea, abdominal pain, and vomiting. Severe symptoms manifest within 2 weeks after establishment as the larval stages invade muscle tissue. These include fever, chills, myalgia (muscle pain), arthralgias (joint pain), and periorbital edema (swelling around the eyes). If a severe infection is not treated, death can result if the heart and diaphragm muscles are affected.

Foodborne illness statistics: *Trichinella* spp. are estimated to cause 156 illnesses, six hospitalizations, and zero deaths in the United States per year (Scallon *et al.*, 2011). Most of these cases were due to

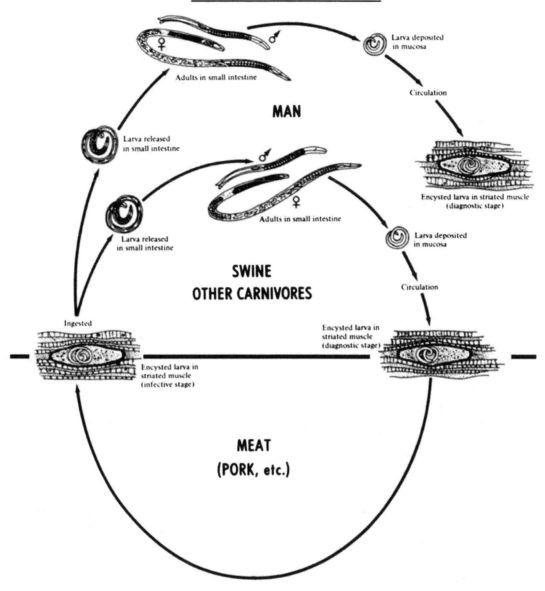

Fig. 8.3. The various stages in the life cycle of *Trichinella spiralis* (CDC PHIL, 2014).

consumption of under-prepared wild game meat. The low incidence in the United States in comparison to other countries can be attributed to successful implementation of food safety practices such as good agricultural practices, meat inspection/testing, and cooking pork to proper internal temperatures.

It is interesting how effective campaigns over the last century have been for control of this parasite in comparison to similar campaigns to control other pathogens that continue to be major sources of foodborne illness, i.e. *Escherichia coli* in ground beef and produce (see Chapter 7).

Costs of foodborne illness: *Trichinella* foodborne infections were estimated to cost US$15,104/case in the United States in 2010 (Scharff, 2012).

Testing methods: Diagnostic images are available from the United States Centers for Disease Control (CDC, 2013). Fayer and Xiao (2015) have described the compression method and the digestion method for testing meat. These methods physically or enzymatically reveal the encysted larvae under microscopic examination of muscle tissue samples.

Controls: Effective controls to lower the risk of foodborne illness from this species include the following:

- The parasite is sensitive to cooking, freezing, and irradiation. Freezing is less consistently effective, especially for meats other than pork (see Box 8.1). The CDC provides information at http://www.cdc.gov/parasites/trichinellosis/prevent.html.
- Education of consumers is essential for how to properly prepare meat to lower the risk of infection.
- Avoiding cross-contamination is important, especially between raw meat and ready to eat foods.

Protozoan foodborne parasites

Protozoa are unicellular eukaryotes (O'Donoghue, 2010). Many of these can cause parasitic infections in humans.

Over recent decades, parasitic protozoa have been recognized as having great potential to cause waterborne and foodborne disease. The organisms of greatest concern in food production worldwide are *Cryptosporidium, Cyclospora, Giardia,* and *Toxoplasma*. Although other parasitic protozoa can be spread by food or water, current epidemiological evidence suggests that these four present the largest risks (Dawson, 2005).

Cryptosporidium

Microscopic morphology: *Cryptosporidium* means hidden spore in Greek. Oocysts are ovular shaped and range in size from 4 to 6 μm (Fig. 8.4).

Environmental factors: *Cryptosporidium* spp. are obligate intracellular parasites. "Sporulated oocysts, containing four sporozoites, are excreted by the infected host through feces and possibly other routes such as respiratory secretions" (CDC, 2014). The oocyte form of the parasite is transmitted through fecal contamination of food or water. Humans or animals can be the source for human infections. Contaminated food can be a carrier, but contaminated water is the primary source for transmission to humans. Swimming pools and undertreated municipal water supplies have been linked most frequently to outbreaks in the United States (see Box 8.2).

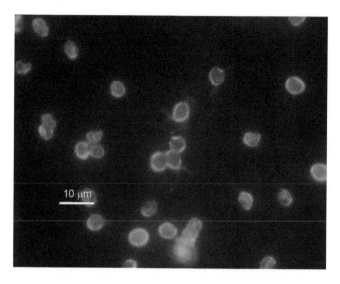

Fig. 8.4. Immunofluorescence image of *Cryptosporidium parvum* oocysts, purified from fecal material (adapted from https://commons. wikimedia.org/wiki/File:Cryptosporidium_ parvum_01.jpg; public domain).

Box 8.2. Crypto takes Milwaukee – 1993

In 1993, over 400,000 residents of Milwaukee, Wisconsin, consumed contaminated municipal water and became infected with *Cryptosporidium parvum*. The cause was a change to the water purification system, which had a flawed filtration system. *Cryptosporidium* oocytes, which are resistant to chemical treatments used in water purification, passed through this defective filtration system and into the water supply (Mac Kenzie *et al.*, 1994).

At this point there is no single, clear source of contamination of the Milwaukee watershed, and we may never know exactly what happened in April 1993. Possible sources include cattle along the two rivers that flow into the Milwaukee harbor above the southern treatment plant, local slaughterhouses and human sewage. Rivers swelled by significant rain and snow runoff may have transported oocysts great distances into the lake and from there to the intake of the southern plant (Blair, 1995).

Immunocompromised populations were especially affected:

During the outbreak, strong associations between turbidity and gastroenteritis-related emergency room visits and hospitalizations occurred at temporal lags of 5–6 days (consistent with the *Cryptosporidium* incubation period). A pronounced second wave of these illnesses in the elderly peaked at 13 days. This wave represented approximately 40% of all excess cases in the elderly. Our findings suggest that the elderly had an increased risk of severe disease due to *Cryptosporidium* infection, with a shorter incubation period than has been previously reported in all adults and with a high risk for secondary person-to-person transmission (Naumova *et al.*, 2003).

In the epidemic, 69 people died, 93% of them patients with AIDS. Others were patients with cancer and children (Behm, 2013).

The total cost of outbreak-associated illness was US$96.2 million: US$31.7 million in medical costs and US$64.6 million in productivity losses. The average total costs for persons with mild, moderate, and severe illness were US$116, US$475, and US$7,808, respectively (Corso *et al.*, 2003).

Type of foodborne illness: *Cryptosporidium parvum* and *C. hominis* are the most common species linked to human parasitic infections. After being consumed, the oocysts excyst and the released sporozoites parasitize intestinal epithelial cells and complete their life cycle. As few as ten oocysts can be an infective dose.

The main symptoms of cryptosporidiosis include self-limiting, but very watery diarrhea, nausea, vomiting,

---

cramps, and fever. Onset time to symptoms ranges from 7 to 10 days. Symptoms typically last about 2 weeks. Hosts can shed oocysts for months after the symptoms clear. The diarrhea can occur in large amounts, and severe dehydration can occur. Cryptosporidiosis can be life threatening for those in immunocompromised states, and can spread to other organ systems.

Foodborne illness statistics: *Cryptosporidium* spp. are estimated to cause 57,616 illnesses, 210 hospitalizations, and four deaths in the United States per year (Scallon *et al.*, 2011).

Costs of foodborne illness: *Cryptosporidium* foodborne infections were estimated to cost US$2,916/case in the United States in 2010 (Scharff, 2012).

Testing methods: Diagnostic images are available from the United States Centers for Disease Control (CDC, 2013). Orlandi *et al.* (2004) have described detection and isolation methods including the polymerase chain reaction and microscopic techniques. Fayer and Xiao (2015) have described tests that include microscopic, enzyme immunoassay kits, rapid immunochromatographic tests, and polymerase chain reaction tests for detecting *Cryptosporidium* in food and environmental samples.

Controls: Effective controls to lower the risk of foodborne illness from this species include the following:

- Use of safe, potable water for food preparation and cleaning is important.
- Good hygiene and handwashing help prevent the spread of this pathogen. *Cryptosporidium* are resistant to chlorine, so chemical sanitizing strategies need to be validated.
- This parasite is sensitive to heating, and boiling of water is often used as an intervention when water treatment systems fail.

Cyclospora cayetanensis

Microscopic morphology: The genus name is derived from the Latin *cycl* and *spora* for circular spore (Fig. 8.5). The species is named after the university where early work resulted in its classification, Cayetano Heredia University in Lima, Peru. This species is a cyst-forming round protozoan. The oocysts are very small, ranging from 7.5 to 10 μm in diameter. Each oocyte contains two sporocysts.

Environmental factors: Humans are the only host for *C. cayetanensis*. Therefore, contamination is from human fecal material. Contaminated water supplies are a primary source for food contamination.

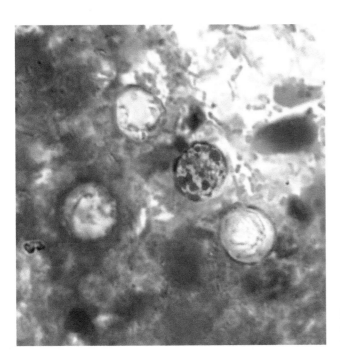

Fig. 8.5. A photomicrograph of *Cyclospora cayetanensis* oocysts in a fresh stool sample (CDC PHIL, 2014; Melanie Moser).

Type of foodborne illness: *C. cayetanensis* causes diarrhea, and is a cause of seasonal and travelers' diarrhea (Herwaldt, 2015). Seasonal diarrhea happens in endemic regions during periods of high precipitation. Endemic regions are in the tropical parts of the world.

Upon ingestion and digestion of the outer coating of the oocysts, the sporocysts invade epithelial cells in the small intestine. The onset time for symptoms to develop ranges from 7 to 10 days. The main symptom is a prolonged, watery diarrhea. Other symptoms can include malaise, low fever, fatigue, vomiting, and weight loss. The duration of symptoms is 3 to 4 days, and the symptoms may recur for up to 4 weeks.

Transmission through contaminated fresh produce occurs in non-endemic regions through international trade. Contamination is a major concern for fresh produce traded internationally.

Foodborne illness statistics: *C. cayetanensis* is estimated to cause 44,407 illnesses, 11 hospitalizations, and zero deaths in the United States per year (Scallon *et al.*, 2011).

Costs of foodborne illness: *C. cayetanensis* foodborne infections were estimated to cost US$1,483/case in the United States in 2010 (Scharff, 2012).

Testing methods: Diagnostic images are available from the United States Centers for Disease Control (CDC, 2013). Orlandi *et al.* (2004) and Fayer and Xiao (2015) have described detection and isolation methods, including the polymerase chain reaction and microscopic techniques.

Controls: Effective controls to lower the risk of foodborne illness from this species include the following:

- Use of safe, potable water for food preparation and cleaning is important.
- Fresh produce should be cleaned thoroughly before consumption.
- Good hygiene and handwashing help prevent the spread of this pathogen. *Cyclospora* are resistant to chlorine, so chemical sanitizing strategies need to be validated.
- This parasite is sensitive to heating, and boiling of water is often used as an intervention when water treatment systems fail.

Giardia

Microscopic morphology: *Giardia* spp. form sesame seed-shaped cysts with distinct flagella (Fig. 8.6). The genus name is based on the discoverer, Alfred Giard. The cysts range in size from 10 to 20 μm in length.

Environmental factors: *Giardia* cysts can survive for up to 3 months in open water, and water contaminated with feces is the primary source of transmission. Humans and other warm-blooded animals can serve as hosts for this parasite. A colloquialism for giardiasis is beaver fever due to its common presence in rivers and streams inhabited with beavers.

Type of foodborne illness: *Giardia intestinalis* is the most common species to infect humans. As with other protozoan parasites, the cysts of *Giardia* are released upon digestion and proceed to invade the

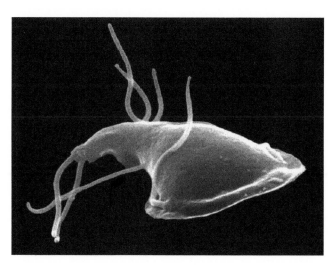

Fig. 8.6. Scanning electron micrograph of a *Giardia* protozoan isolated from a rat's intestine (CDC PHIL, 2014; Stan Erlandsen and Dennis Feely).

Eukaryotic Microorganisms of Concern in Food

epithelial lining of the small intestine. An infectious dose can be one cyst.

Symptoms include foul-smelling diarrhea, cramps, fatigue, weight loss, fever, and vomiting. The onset of symptoms ranges from 1 to 2 weeks. The duration of symptoms can be 2 to 6 weeks, and in a few cases can last years. Asymptomatic carrier states are common. The most severe cases can result in death due to complications of dehydration and malabsorption of nutrients.

Giardia is the most common cause of human parasitic infections worldwide, and the second most common cause of confirmed foodborne parasitic infections in the United States.

> *Giardia* is endemic worldwide. *Giardia* was the most commonly diagnosed pathogen in travelers seeking medical attention for a gastrointestinal infection at GeoSentinel surveillance clinics worldwide after returning from the Caribbean, Central and South America, Western Europe, North Africa, sub-Saharan Africa, the Middle East, South Asia, and East Asia and was the second most commonly diagnosed pathogen in travelers returning from Southeast Asia and Australasia. The risk of infection increases with duration of travel; backpackers or campers who drink untreated water from lakes or rivers are also more likely to be infected. *Giardia* infections are commonly identified in internationally adopted children, although many are asymptomatic (Gargano and Yoder, 2015).

Foodborne illness statistics: *G. intestinalis* is estimated to cause 76,840 illnesses, 225 hospitalizations, and two deaths in the United States per year (Scallon *et al.*, 2011).

Costs of foodborne illness: *G. intestinalis* foodborne infections were estimated to cost US$3,672/case in the United States in 2010 (Scharff, 2012).

Testing methods: Diagnostic images are available from the United States Centers for Disease Control (CDC, 2013). Fayer and Xiao (2015) have described microscopic and enzyme immunoassay methods for detection.

Controls: Effective controls to lower the risk of foodborne illness from this species include the following:

- Use of safe, potable water for food preparation and cleaning is important.
- Fresh produce should be cleaned thoroughly before consumption.
- Good hygiene and handwashing help prevent the spread of this pathogen. *Cyclospora* are resistant

to chlorine, so chemical sanitizing strategies need to be validated.

- This parasite is sensitive to heating, and boiling of water is often used as an intervention when water treatment systems fail.

Toxoplasma gondii

Microscopic morphology: *T. gondii* forms round (coccidian) infectious oocysts. The genus name is derived from the Greek word *toxon* describing the arc-like shape of the excysted forms (tachyzoites). The species name is based on the host species from which it was originally identified, the gundi, a small rodent found in Africa.

Environmental factors: The life cycle of *T. gondii* (Fig. 8.7) is more complex than those of other foodborne protozoan parasites. It requires both an intermediate host and a definitive host. The intermediate hosts support the asexual form, and hosts can include a wide variety of warm-blooded animals, including humans. The definitive host supports the sexual form of the parasite, and these hosts are felines. Infected felines shed oocysts into the environment to infect the next intermediate host. Humans become infected by consuming food contaminated with feline fecal material or consumption of infective intermediate forms (bradyzoites) in contaminated meat that is raw or undercooked. Oocysts can survive environmental conditions outside of the host for long periods. The infectivity remains intact even after long periods outside of a host.

Type of foodborne illness: *T. gondii* causes a parasitic infection. This is the most common, most deadly, and most expensive type of foodborne parasitic infection in the United States. It is the most common human parasite in developed countries, but is endemic worldwide (Jones, 2015).

The infective oocyst invades the small intestine. The cysts are capable of invading additional tissues, and can cross barriers to the brain and cross the placenta to the fetus.

Symptoms are similar to those of mononucleosis, and include fever, headache, sore lymph nodes, myalgia, and rash. Onset time for symptoms ranges from 5 to 23 days. In severe cases, additional symptoms may include confusion, nausea, poor coordination, and seizures. In cases where the infection can affect vision, symptoms may include blurred vision, tearing, redness, and eye pain. Chronic infection

Toxoplasmosis

(Toxoplasma gondii)

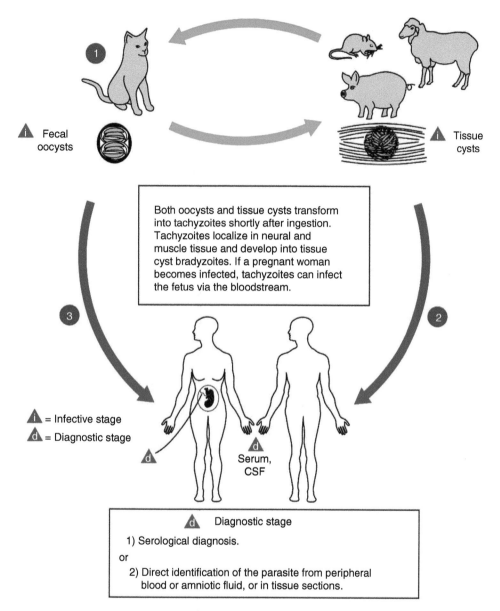

Fecal oocysts

Tissue cysts

Both oocysts and tissue cysts transform into tachyzoites shortly after ingestion. Tachyzoites localize in neural and muscle tissue and develop into tissue cyst bradyzoites. If a pregnant woman becomes infected, tachyzoites can infect the fetus via the bloodstream.

🔺 = Infective stage
🔺 = Diagnostic stage

Serum, CSF

🔺 Diagnostic stage

1) Serological diagnosis.

or

2) Direct identification of the parasite from peripheral blood or amniotic fluid, or in tissue sections.

Fig. 8.7. Illustration of the life cycle of *Toxoplasma gondii* (CDC PHIL, 2014; Alexander J. da Silva and Melanie Moser).

may result in cysts invading muscle, heart, brain, and skeletal tissues. Asymptomatic carrier states are common.

Infection in pregnant women can cause birth defects for fetuses. In children, infection may result in central nervous system disorders including mental retardation and visual impairment.

Foodborne illness statistics: *T. gondii* is estimated to cause 86,686 illnesses, 4,428 hospitalizations, and 327 deaths in the United States per year (Scallon *et al.*, 2011). This is the second most deadly foodborne pathogen in confirmed cases in the United States.

Costs of foodborne illness: *Toxoplasma gondii* foodborne infections were estimated to cost US$39,869/case in the United States in 2010 (Scharff, 2012).

Testing methods: Diagnostic images are available from the United States Centers for Disease Control (CDC, 2013). Fayer and Xiao (2015) have described biological assays for detection.

Controls: Effective controls to lower the risk of foodborne illness from this species include the following:

- Pregnant women should avoid cleaning cat litter boxes.
- Use of safe, potable water for food preparation and cleaning is important.
- Fresh produce should be cleaned thoroughly before consumption.
- Good hygiene and handwashing help prevent the spread of this pathogen. *Toxoplasma* are resistant to chlorine, so chemical sanitizing strategies need to be validated.
- The parasite is sensitive to cooking, freezing, and irradiation.
- Education of consumers on how to properly prepare meat to lower the risk of infection is essential.
- Avoiding cross-contamination is important, especially between raw meat and ready-to-eat foods.

Molds and Mycotoxins

Molds are filamentous fungi. Fungi are their own biological kingdom that includes mushrooms, yeasts, and molds. Fungi can be nutritious food, but many are the human race's biggest competition for our food supply. If a substance is food for a human, it is also food for fungi. Mold exists in almost any environment on the planet, and spores are present in the air. It is therefore quite likely that food at any stage is inoculated with fungal spores waiting for the right environmental conditions to grow. Many mold species can grow at lower water activities than bacteria and become spoilage and safety problems in low-moisture food commodities.

Many molds have evolved to produce potent toxins that give them a competitive advantage for procuring food. The toxins produced by molds are

called mycotoxins, and thousands of these have been identified. Mycotoxins are secondary metabolites (see Chapter 2). Most mycotoxins are heat stable and remain biologically active after food processing. As with all toxins, the toxicity is dependent upon the dose consumed (see Chapter 5). Humans likely often consume low levels of mycotoxins, and zero tolerances for any level of mycotoxin would result in food shortages and dramatic economic impacts. Many mycotoxins are regulated to control risk levels (Murphy *et al.*, 2006). Mycotoxin contamination of food and feed materials has global impacts on food safety and security (Cardwell *et al.*, 2001).

The fungi of most concern in food safety fall into the categories of field fungi and storage fungi.

Field fungi are the plant pathogens that invade food crops as they form. Storage fungi are those saprophytic fungi that manage to grow in stored food and feed materials. The three genera of most concern in food safety are *Aspergillus*, *Fusarium*, and *Penicillium*. The chemical structure of the major mycotoxins of most concern can be viewed online via Murphy *et al.* (2006).

Aspergillus

Microscopic morphology: *Aspergillus* spp. display typical growth patterns on various media. *Aspergillus* can be described as a "Hyaline hyphomycete showing distinctive conidial heads with flask-shaped phialides arranged in whorls on a vesicle" (Mycology Online, 2016; also see Figure 8.8). The genus name is derived from the Latin word *aspergillium*, which describes the device used in churches for sprinkling holy water.

Environmental factors: *Aspergillus* can be both field and storage fungi, and many verge on being xerophylic (dry loving; see Chapter 3) in nature. In drought conditions, some *Aspergillus* can invade crops such as peanuts and corn in the field. The risk of *Aspergillus* producing mycotoxins is highest in stored material, and water activities as low as 0.8 can support growth and mycotoxin production.

Type of foodborne illness: Mycotoxicoses caused by *Aspergillus* can be due to several forms of mycotoxins produced by various species within this genus. Many of these are of food safety concern; however, the main species of concern are *Aspergillus flavus* and *A. parasiticus*. These species produce potent mycotoxins called aflatoxins.

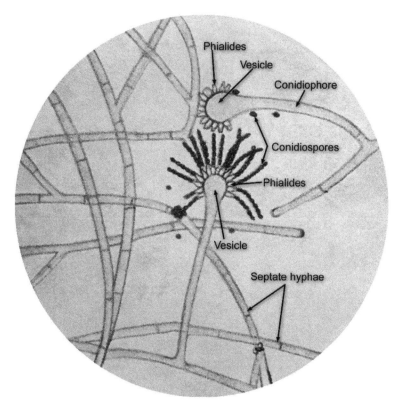

Fig. 8.8. An illustration of *Aspergillus* mold with septate hyphae, conidiophores, which support the apparatus responsible for the development of the organism's asexual conidiophores, i.e. vesicle, and phialides. Magnified 125× (CDC PHIL, 2014).

Aflatoxins can cause acute intoxications in the form of organ damage, particularly liver damage. They more commonly cause chronic illness such as cancer. Aflatoxins are enzymatically transformed by liver enzymes to reactive forms that interact with proteins and DNA; hence the liver is the organ of primary exposure. Metabolites of aflatoxins can be transmitted through milk, making regulation of dairy animal feed a necessity to prevent human exposure.

Because aflatoxins are one of the most significant risk factors for liver cancer, one of the deadliest cancers worldwide, controlling its presence in the food supply is critical. It is possibly responsible for up to 172,000 liver cancer cases per year, most of which would result in mortality within three months of diagnosis. Possibly even more critical from a global public health standpoint is the link between aflatoxin exposure and childhood stunting, which can lead to a variety of adverse health conditions that last well beyond childhood. However, at the moment there is insufficient evidence for a quantitative risk assessment to evaluate exact daily doses of aflatoxins that lead to particular levels of risk or adverse health outcomes in children. Additionally, while aflatoxins may lead to immunomodulation, not enough information is currently known about how this leads to particular adverse health outcomes in humans. However, the human health evidence points to aflatoxins' association with multiple adverse effects; hence, it is important to reduce human exposures to aflatoxins in the diet to the extent that feasible methods allow (Wu, 2013).

Foodborne illness statistics: Mycotoxicoses are rare to non-existent in the United States due to regulatory practices. Chronic and acute illnesses are more common in developing countries. Acute illnesses are relatively rare in human populations, and occur when food supplies are severely limited to moldy grain. Chronic illnesses, such as cancer, are difficult to monitor as causation is not always clear where cancer rates are higher than normal.

Costs of foodborne illness: Specific costs on per case bases are not available in the literature for intoxications due to *Aspergillus* mycotoxins. See Box 8.3 for more information.

Testing methods: Tournas *et al.* (2001) and Gourama *et al.* (2015) describe various analytical and rapid testing methods for mycotoxin testing. The United States Department of Agriculture provides testing and sampling information through the Grain Inspection, Packers and Stockyards Program (GIPSA, 2016). The European Mycotoxin Awareness Network provides similar information (EMAN, 2015).

Controls: Effective controls to lower the risk of foodborne illness from this species include the following:

- Keeping dry materials from reaching water activities that allow mold growth is the most effective control for storage fungi producing mycotoxins.
- Purchasing materials from reputable sources is recommended.
- Exclusion is a common practice to try to divert contaminated material away from the human food supply. This approach can have unintended consequences (see Box 8.3).
- Physical, chemical, and biological control techniques can sometimes reduce levels to acceptable

limits. These types of approaches require careful and consistent validation.

Fusarium

Microscopic morphology: The genus name is derived from the Latin word *fusus* meaning spindle. This describes the shape of macroconidia that some species produce. "The character which defines the genus *Fusarium* is the production of septate, fusiform to crescent-shaped conidia, termed macroconidia, with a foot-shaped basal cell and a more or less beaked apical cell" (Pitt and Hocking, 2009; also see Fig. 8.9).

Environmental factors: *Fusarium* spp. are typically considered a field fungi, but can be storage molds if moisture contents are high enough. Many *Fusarium* are considered plant pathogens. *Fusarium graminearum* is a common cause of scab, or *Fusarium* head blight, in grasses such as wheat. This plant disease results in mycotoxin contamination of the seeds as well as overall loss in grain quality and yield.

Fusarium plant disease epidemics in crops tend to occur during growing seasons where there are extended periods of high moisture or relative humidity and moderately warm temperatures (Leonard and Bushnell, 2003; Schmale and Bergstrom, 2003).

Box 8.3. Unintended consequences of aflatoxin regulation

An old industry saying is "dilution is the solution for the pollution". In practice, what this means is that food or feed material that is above tolerance concentrations for contaminants can be mixed with higher quality materials to dilute the contamination to acceptable levels. This can be a practical and economically sound way to manage financial and food safety risks. However, strict regulations for controlling concentrations of aflatoxins in food and feed limit the possibilities, and can result in unintended consequences for vulnerable human and animal populations.

Aflatoxins are highly regulated in many countries. For countries that produce food for export, but that do not as closely regulate domestic supplies, this can have disastrous effects. Regulations appear to drive trade patterns, resulting in more contaminated material being traded in clusters of countries with lower standards (Wu and Guclu, 2012). This all has

an effect of concentrating contaminated materials in regions that cannot export material exceeding threshold limits for aflatoxins, and that tend to have inherent geographically related environmental conditions that put stored crops at higher risk of contamination.

Kenya is such a country and has experienced multiple outbreaks of acute aflatoxicosis resulting in human mortality.

Kenya is one of the world's hotspots for aflatoxins, with what is believed to be the highest incidence of acute toxicity ever documented. This country suffered severe outbreaks of illness from aflatoxins in 2004 and 2010, poisoning more than 300 people in the 2004 event alone, and killing more than 100 of them. Domestic animals that consume feeds contaminated by aflatoxins also can become sick and die (Macmillan, 2014).

Fig. 8.9. A microscopic image of a *Fusarium* mold with septate hyphae, and the floral arrangement of the fusiform-shaped macroconidia. Magnified 475× (from CDC PHIL, 2014).

Type of foodborne illness: Mycotoxicoses caused by *Fusarium* can be due to many mycotoxigenic species within this genus.

Trichotecenes is a family of over 100 structurally related and naturally occurring mycotoxins produced by *Fusarium*. Of these, T-2 toxin is the most potent, and deoxynivalenol (DON, also known as vomitoxin) is the most common and is abundantly produced in food crops. Both of these toxins may cause acute intoxications and chronic health issues and are of major agro-economic concern. T-2 toxin causes an acute intoxication that takes from 1 to 12 hours for symptoms to manifest. Symptoms may include blisters, necrosis of tissues, dizziness, nausea, vomiting, diarrhea, hemorrhaging, hemolysis of erythrocytes, and death. DON causes nausea, vomiting, and feed refusal in animals.

Another major category of mycotoxins produced by *Fusarium* includes the fumonisins. *Fusarium verticillioides* (formerly *moniliforme*) and *F. proliferatum* are the main species of concern in grain safety. "Fumonisins can cause leukoencephalomalacia in horses and pulmonary edema in pigs and have been linked to a variety of significant adverse health effects in other livestock and experimental animals" (FDA, 2014).

Foodborne illness statistics: Mycotoxicoses are rare to non-existent in the United States due to regulatory practices. Chronic and acute illnesses are more common in developing countries. Acute illnesses are relatively rare in human populations, and occur when food supplies are severely limited to moldy grain.

Costs of foodborne illness: Specific costs on per case bases are not available in the literature for intoxications due to *Fusarium* mycotoxins.

Testing methods: Tournas *et al.* (2001) and Gourama *et al.* (2015) have described various analytical and rapid testing methods for mycotoxin testing. The United States Department of Agriculture provides testing and sampling information through the Grain Inspection, Packers and Stockyards Program (GIPSA, 2016). The European Mycotoxin Awareness Network provides similar information (EMAN, 2015).

Controls: Effective controls to lower the risk of foodborne illness from this species include the following:

- Purchasing materials from reputable sources is recommended.
- Exclusion is a common practice to try to divert contaminated material away from the human food supply. This approach can have unintended consequences (see Box 8.3).
- Dilution may be allowed to decrease concentrations of *Fusarium* mycotoxins.
- Physical, chemical, and biological control techniques can sometimes reduce levels to acceptable limits. These types of approaches require careful and consistent validation.

Penicillium

Microscopic morphology: The genus name is derived from the Latin word *penicillus*, which means paintbrush, describing the appearance of the conidiophores.

"Microscopically, chains of single-celled conidia (ameroconidia) are produced in basipetal succession from a specialized conidiogenous cell called a phialide" (Mycology Online, 2016; also see Fig. 8.10).

Environmental factors: *Penicillium* spp. are typically considered to be a storage fungi. Many *Penicillium* are considered psychrotrophs (see Chapter 3) and can grow at refrigeration temperatures.

Type of foodborne illness: Mycotoxicoses caused by *Penicillium* can be due to many mycotoxigenic species within this genus. The two most regulated mycotoxins produced by this genus include ochratoxin and patulin. Both have shown the potential to be carcinogenic in laboratory studies.

Foodborne illness statistics: Mycotoxicoses are rare to non-existent in the United States due to regulatory practices.

Costs of foodborne illness: Specific costs on per case bases are not available in the literature for intoxications due to *Penicillium* mycotoxins.

Testing methods: Tournas *et al.* (2001) and Gourama *et al.* (2015) describe various analytical and rapid testing methods for mycotoxin testing. The United States Department of Agriculture provides testing and sampling information through the Grain Inspection, Packers and Stockyards Program (GIPSA, 2016). The European Mycotoxin Awareness Network provides similar information (EMAN, 2015).

Controls: Effective controls to lower the risk of foodborne illness from this species include the following:

- Limiting time in storage for foods commonly contaminated with *Penicillium* molds prevents growth and mycotoxin production.
- Purchasing materials from reputable sources is recommended.

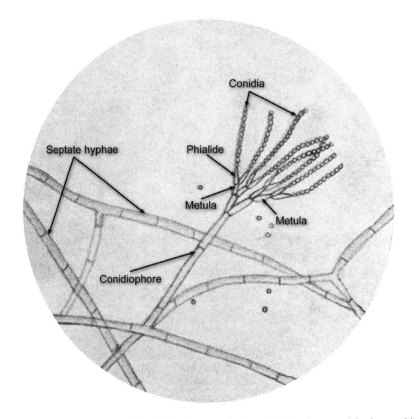

Fig. 8.10. An illustration of *Penicillium* mold including the organism's septate hyphae, conidiophore, which support the apparatus responsible for the development of the organism's asexual conidiophores, i.e. metulae and phialides. Magnified 125× (CDC PHIL, 2014).

- Exclusion is a common practice to try to divert contaminated material away from the human food supply. This approach can have unintended consequences (see Box 8.3).
- Dilution may be allowed to decrease concentrations of *Penicillium* mycotoxins.
- Physical, chemical, and biological control techniques can sometimes reduce levels to acceptable limits. These types of approaches require careful and consistent validation.

Summary

The eukaryotic microorganisms from the animal and fungal biological kingdoms that present the most concern for food safety have been described. The parasites discussed include helminths and protozoa. The mycotoxins of most concern for food safety have also been described. The parasites cause parasitic infections, and the molds of concern cause intoxications through the production of mycotoxins that are resistant to food processing techniques.

Further Reading

FDA (2012) *Bad Bug Book*, 2nd edn. United States Food and Drug Administration, Washington, DC. Available at: http://www.fda.gov/Food/FoodborneIllnessContaminants/CausesOfIllnessBadBugBook/default.htm (accessed 29 December 2015).

Juneja, V.K. and Sofos, J.N. (2010) *Pathogens and Toxins in Foods: Challenges and Interventions*. ASM Press, Washington, DC.

References

Audicana, M.T. and Kennedy, M.W. (2008) *Anisakis simplex*: from obscure infectious worm to inducer of immune hypersensitivity. *Clinical Microbiology Reviews* 21, 360–379.

Behm, D. (2013) Milwaukee marks 20 years since *Cryptosporidium* outbreak. *Journal Sentinel*. Available at: http://www.jsonline.com/news/milwaukee/milwaukee-marks-20-years-since-cryptosporidium-outbreak-099dio5-201783191.html (accessed 29 December 2015).

Bier, J.W., Jackson, G.J., Adams, A.M., and Rude, R.A. (2001) Chapter 19. Parasitic animals in foods. In: *Bacteriological Analytical Manual (BAM)*. United States Food and Drug Administration, Washington, DC. Available at: http://www.fda.gov/Food/FoodScienceResearch/LaboratoryMethods/ucm2006949.htm (accessed 29 December 2015).

Blair, K. (1995) *Cryptosporidium* and public health. *Water Quality and Health*. Available at: http://www.waterandhealth.org/newsletter/old/03-01-1995.html. (accessed 29 December 2015).

Cantey, P.T. and Jones, J.L. (2015) Chapter 3. Infectious diseases related to travel: taeniasis. In: *The Yellow Book: CDC Health Information for International Travel 2016*. United States Centers for Disease Control and Prevention, Washington, DC. Available at: http://wwwnc.cdc.gov/travel/page/yellowbook-home-2014 (accessed 29 December 2015).

Cardwell, K.F., Desjardins, A., Henry, S.H., Munkvold, G., and Robens, J. (2001) Mycotoxins: the cost of achieving food security and food quality. *APSnet Features*. Available at: http://www.apsnet.org/publications/apsnetfeatures/Pages/Mycotoxins.aspx (accessed 29 December 2015).

CDC (2004) Trichinellosis associated with bear meat – New York and Tennessee, 2003. *Morbidity and Mortality Weekly Report* 53, 606–610.

CDC (2013) Laboratory identification of parasitic diseases of public health concern. Available at: http://www.cdc.gov/dpdx/az.html (accessed 29 December 2015).

CDC (2014) About parasites. Available at: http://www.cdc.gov/parasites/about.html (accessed 29 December 2015).

CDC PHIL (2014) United States Centers for Disease Control Public Health Image Library. Available at: http://phil.cdc.gov/phil/home.asp (accessed 29 December 2015).

Corso, P.S., Kramer, M.H., Blair, K.A., Addiss, D.G., Davis, J.P., and Haddix, A.C. (2003) Cost of illness in the 1993 waterborne *Cryptosporidium* outbreak, Milwaukee, Wisconsin. *Emerging Infectious Diseases* 9, 426–431.

Dawson, D. (2005) Foodborne protozoan parasites. *International Journal of Food Microbiology* 103, 207–227.

EMAN (2015) European mycotoxins awareness network. Available at: http://eman.leatherheadfood.com/ (accessed 29 December 2015).

Fayer, R. and Xiao, L. (2015) Chapter 42. Waterborne and foodborne parasites. In: Salfinger, Y. and Tortorello, M.L. (eds) *Compendium of Methods for the Microbiological Examination of Foods*. Association of Public Health Laboratories, Washington, DC, pp. 565–582.

FDA (2001) Chapter V. Potential hazards in cold-smoked fish: parasites. In: *Processing Parameters Needed to Control Pathogens in Cold Smoked Fish*. United States Food and Drug Administration, Washington, DC. Available at: http://www.fda.gov/Food/FoodScienceResearch/SafePracticesforFoodProcesses/ucm094578.htm (accessed 29 December 2015).

FDA (2014) Fumonisins. Available at: http://www.fda.gov/AnimalVeterinary/Products/AnimalFoodFeeds/Contaminants/ucm050311.htm (accessed 29 December 2015).

Gargano, J.W. and Yoder, J.S. (2015) Chapter 3. Infectious diseases related to travel: giardiasis. In: *The Yellow Book: CDC Health Information for International Travel 2016*. United States Centers for Disease Control and Prevention, Washington, DC. Available at: http://wwwnc.cdc.gov/travel/page/yellowbook-home-2014 (accessed 29 December 2015).

GIPSA (2016) Mycotoxins. United States Department of Agriculture: grain inspection, packers and stockyards administration. Available at: https://www.gipsa.usda.gov/fgis/mycotoxins.aspx (accessed 1 March 2016).

Gourama, H., Bullerman, L.B., and Bianchini, A. (2015) Chapter 43. Toxigenic fungi and fungal toxins. In: Salfinger, Y. and Tortorello, M.L. (eds) *Compendium of Methods for the Microbiological Examination of Foods*. Association of Public Health Laboratories, Washington, DC, pp. 583–594.

Herwaldt, B.L. (2015) Chapter 3. Infectious diseases related to travel: cyclosporiasis. In: *The Yellow Book: CDC Health Information for International Travel 2016*. United States Centers for Disease Control and Prevention, Washington, DC. Available at: http://wwwnc.cdc.gov/travel/page/yellowbook-home-2014 (accessed 29 December 2015).

Hochberg, N.S. and Hamer, D.H. (2010) Anisakidosis: perils of the deep. *Emerging Infections* 51, 806–812.

Ito, A. and Craig, P.S. (2003) Immunodiagnosis and molecular approaches for the detection of taeniid cestode infections. *Trends in Parasitology* 19, 377–381.

Jones, J.L. (2015) Chapter 3. Infectious diseases related to travel: toxoplasmosis. In: *The Yellow Book: CDC Health Information for International Travel 2016*. United States Centers for Disease Control and Prevention, Washington, DC. Available at: http://wwwnc.cdc.gov/travel/page/yellowbook-home-2014 (accessed 29 December 2015).

Leonard, K.J. and Bushnell, W.R. (2003) *Fusarium Head Blight of Wheat and Barley*. American Phytopathological Society, St Paul, MN, USA.

Mac Kenzie, W.R., Hoxie, N.J., Proctor, M.E., Gradus, M.S., Blair, K.A., Peterson, D.E., Kazmierczak, J.J., Addiss, D.G., Fox, K.R., Rose, J.B., and Davis, J.P. (1994) A massive outbreak in Milwaukee of *Cryptosporidium* infection transmitted through the public water supply. *New England Journal of Medicine* 331, 161–167.

Macmillan, S. (2014) Aflatoxins in Kenya's food chain: overview of what researchers are doing to combat the threat to public health. *ILRI News*. Available at: https://news.ilri.org/2014/05/06/aflatoxins-in-kens-food-chain/ (accessed 11 October 2016).

Murphy, P., Hendrich, S., Landgren, C., and Bryant, C.B. (2006) Food mycotoxins: an update. *Journal of Food Science* 71, 51–65. Available at: http://www.ift.org/knowledge-center/read-ift-publications/science-reports/scientific-status-summaries/food-mycotoxins.aspx (accessed 29 December 2015).

Mycology Online (2016) National Mycology Reference Center. University of Adelaide. Available at: http://www.mycology.adelaide.edu.au/ (accessed 1 March 2016).

Naumova, E.N., Egorov, A.I., Morris, R.D., and Griffiths, J.K. (2003) The elderly and waterborne *Cryptosporidium* infection: gastroenteritis hospitalizations before and during the 1993 Milwaukee outbreak. *Emerging Infectious Diseases* 9, 418–425.

NOAA (2007) Harmful algae. National Office for Harmful Algal Blooms; National Oceanic and Atmospheric Administration Center for Sponsored Coastal Ocean Research. Available at: http://www.whoi.edu/redtide/ (accessed 29 December 2015).

O'Donoghue, P. (2010) PARA-SITE. Available at: http://parasite.org.au/para-site/introduction/index.html (accessed 29 December 2015).

Orlandi, P.A., Chu, D.M.T., Bier, J.W., and Jackson, G.J. (2002) Parasites and the food supply. *Food Technology* 56, 72–81.

Orlandi, P.A., Frazar, C., Carter, L., and Chu, D.M.T. (2004) Chapter 19A. Detection of *Cyclospora* and *Cryptosporidium* from fresh produce: isolation and identification by polymerase chain reaction (PCR) and microscopic analysis. In: *Bacteriological Analytical Manual (BAM)*. United States Food and Drug Administration, Washington, DC. Available at: http://www.fda.gov/Food/FoodScienceResearch/LaboratoryMethods/ucm2006949.htm (accessed 29 December 2015).

Pitt, J.I. and Hocking, A.D. (2009) *Fungi and Food Spoilage*, 3rd edn. Springer, New York, USA.

Scallon, E., Hoekstra, R.M., Angulo, F.J., Tauxe, R.V., Widdowson, M.A., Roy, S.L., Jones, J.L., and Griffin, P.M. (2011) Foodborne illness acquired in the United States – major pathogens. *Emerging Infectious Diseases* 17, 7–15.

Scharff, R.L. (2012) Economic burden from health losses due to foodborne illness in the United States. *Journal of Food Protection* 75, 123–131.

Schmale, D.G. and Bergstrom, G.C. (2003) *Fusarium* head blight in wheat. *The Plant Health Instructor*. doi:10.1094/PHI-I-2003-0612-01.

Tournas, V., Stack, M.E., Mislivec, P.B., Koch, H.A., and Bandler, R. (2001) Chapter 18. Yeasts, molds and mycotoxins. In: *Bacteriological Analytical Manual (BAM)*. United States Food and Drug Administration, Washington, DC. Available at: http://www.fda.gov/Food/FoodScienceResearch/LaboratoryMethods/ucm2006949.htm (accessed 29 December 2015).

WHO (2016) Intestinal worms. Available at: http://www.who.int/intestinal_worms/en/ (accessed 10 January 2016).

Wu, F. (2013) Aflatoxins: finding solutions for improved food safety; aflatoxin exposure and chronic human diseases: estimates of burden of disease. Available at: https://news.ilri.org/2014/05/06/aflatoxins-in-kens-food-chain/ (accessed 29 December 2015).

Wu, F. and Guclu, H. (2012) Aflatoxin regulations in a network of global maize trade. *PLoS ONE* 7, e45151.

9 Viruses and Prions

Key Questions
- What is a virus?
- What is a prion?
- What viruses and prions are of most concern for microbial food safety?
- What controls can prevent foodborne illness due to viruses and prions?

Not Alive?

In biological terms, neither viruses nor prions are considered independent life forms. Both have biological activity and can replicate. They are similar to other parasites in the sense that they require a living host, but differ in that they rely on the biochemical machinery of a host to replicate. They can be considered to be on the border between chemistry and life (Villarreal, 2008; Halfmann, 2014). Both are infectious agents and of concern for food safety.

Viruses

"A virus is basically a tiny bundle of genetic material—either DNA or RNA—carried in a shell called the viral coat, or capsid, which is made up of bits of protein called capsomeres" (ASM, 2014). Virus particles are very small (can be 10,000 times smaller than bacteria) and require an electron microscope to reach the magnification needed to be able to see them.

The viruses of most concern in food safety include noroviruses, hepatitis A virus, and rotaviruses. These are considered enteroviruses.

Hepatitis A virus

Microscopic morphology: Hepatitis A virus (HAV) is a single-stranded, positive-sense RNA virus with a very small, icosahedral symmetry (20 sides), non-enveloped structure. HAV particles are about 30 nm in diameter (Fig. 9.1). They are composed of only protein and RNA (WHO, 2016).

Environmental factors: Any ready-to-eat food can be a transmission source if contaminated with virus particles. Virus particles are passed from infected hosts in fecal material. Shellfish are a common food source as they can concentrate the HAV particles out of the water by filter feeding in polluted aquatic environments.

Type of foodborne illness: The viral infection caused by HAV only occurs in humans and other primates. The viruses are transmitted via the fecal–oral route. An infectious dose can be as few as ten HAV particles. Symptoms include inflammation of the liver, fever, nausea, abdominal pain, fatigue, and jaundice. The onset time from exposure to symptoms can range from 15 to 150 days. Hosts are most infectious 1 to 2 weeks prior to onset of symptoms. Most HAV infections are self-limiting. The duration of acute symptoms lasts from 1 week to 6 months. In rare cases, liver damage can become severe and life threatening. Because of the long periods of shedding of virus particles, especially for asymptomatic carriers, epidemics can be challenging to contain.

Foodborne illness statistics: Foodborne HAV causes 1,566 illnesses, 99 hospitalizations, and seven deaths per year in the United States (Scallon *et al.*, 2011).

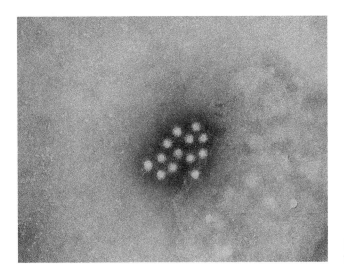

Fig. 9.1. Electron micrograph of hepatitis A virus particles (CDC PHIL, 2014; Betty Partin).

Africa and Asia are highly endemic areas (Nelson and Murphy, 2015).

Costs of foodborne illness: HAV infections were estimated to cost US$37,073/case in the United States in 2010 (Scharff, 2012).

Testing methods: Richards *et al.* (2015) and Woods (2014) have described the molecular methods for HAV detection in foods. Particles typically need to be concentrated to reach detection limits.

Controls: Effective controls to lower the risk of foodborne illness include the following:

- Good hygiene and sanitation practices prevent fecal contamination. HAV particles are resistant to chemical sanitizers, so physical removal through cleaning is needed.
- Avoid raw seafood.
- The virus particles are sensitive to cooking. A temperature of 88°C (190°F) for a few minutes is needed for inactivation.
- Keep food handlers who are ill away from food preparation areas.
- A vaccine is available to prevent HAV infection, and can be an effective control measure for food handlers.

Noroviruses

Microscopic morphology: Noroviruses, formerly known as Norwalk viruses, are small, non-enveloped, icosahedral, single-stranded, positive-sense RNA viruses (Fig. 9.2). Their diameter ranges from 23 to 40 nm. This virus type was first isolated and identified from a 1968 outbreak in a school in Norwalk, Ohio.

Environmental factors: Any ready-to-eat food can be a transmission source if contaminated with virus particles.

Type of foodborne illness: Norovirus causes a highly contagious viral infection. An infectious dose can be as few as 18 particles. See Box 9.1 for an example of how this virus can cause a large-scale outbreak. The virus invades host cells in the small and large intestines, but the molecular mechanisms of pathogenesis are not well understood. The onset time to symptoms is 12 to 48 hours. The main symptoms include vomiting and diarrhea, with other possible symptoms of abdominal cramps, nausea, and fever. Symptoms last 12 to 60 hours. Asymptomatic carrier states can be common.

> Transmission occurs primarily through the fecal–oral route, either through direct person-to-person contact or indirectly via contaminated food or water. Norovirus is also spread through aerosols of vomitus and contaminated environmental surfaces and objects.... Large norovirus outbreaks are associated with settings where people live in close quarters and can easily infect each other, such as hotels, cruise ships, camps, dormitories, and hospitals. Viral contamination of inanimate objects or environmental surfaces (fomites) may persist during and after outbreaks and be a source of infection. On cruise ships, for instance, such environmental contamination has caused recurrent

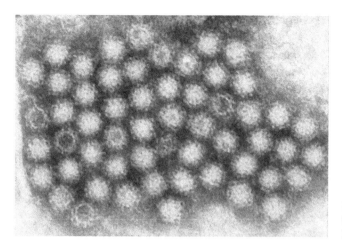

Fig. 9.2. Transmission electron micrograph of norovirus particles (CDC PHIL, 2014; Charles D. Humphrey).

Box 9.1. The best bakery in Minnesota has just had an accident

"The Best Bakery In Minnesota Has Just Had an Accident" was the headline in a newspaper for a story about a foodborne illness outbreak that occurred in Minnesota. This outbreak occurred from 23 to 26 August 1982 and included an estimated 3000 cases. The vehicle of transmission was determined to be contaminated cake frosting. "Observation of the actual procedure showed that the person preparing the frosting often submerged his bare arm up to the elbow in the frosting as it was being mixed to break sugar lumps and scrape the sides of the vat" (Kuritsky *et al.*, 1984).

This large-scale outbreak demonstrates how broadly and quickly highly contagious norovirus can spread. Regardless of the outbreak, the bakery in question actually garnered more business after this outbreak as consumers seemed to tune out the outbreak and hone in on the part about it being "the best bakery in Minnesota."

---

norovirus outbreaks on successive cruises with newly boarded passengers. Transmission of norovirus on airplanes has been reported during both domestic and international flights and likely results from contamination of lavatories or from symptomatic passengers in the cabin (Hall and Lopman, 2015).

Foodborne illness statistics: Foodborne norovirus causes 5,461,731 illnesses, 14,663 hospitalizations, and 149 deaths per year in the United States (Scallon *et al.*, 2011). Norovirus is the leading cause of verified microbial-caused foodborne illnesses in the United States.

Norovirus is ubiquitous, associated with 18% (95% CI: 17-20%) of diarrheal disease globally, with similar proportions of disease in high- middle- and low-income settings. Norovirus is estimated to cause approximately 200,000 deaths annually worldwide, with 70,000 or more among children in developing countries. The entire age range is affected, with children experiencing the highest incidence. Severe outcomes, including hospitalization and deaths, are common among children and the elderly. In both high- and middle-income countries with mature rotavirus vaccination programs, norovirus is being unmasked as the most common cause of pediatric gastroenteritis requiring medical care (CDC, 2015a).

Costs of foodborne illness: Norovirus infections were estimated to cost US$673/case in the United States in 2010 (Scharff, 2012).

Testing methods: Richards *et al.* (2015) have described molecular methods for norovirus detection in foods. Particles typically needed to be concentrated to reach detection limits.

Controls: Effective controls to lower the risk of food-borne illness from this species include the following:

- Good hygiene and sanitation practices prevent contamination.
- The virus particles are sensitive to cooking.
- Keep food handlers who are ill away from food preparation areas.

Rotavirus

Microscopic morphology: Rotaviruses are non-enveloped, double-stranded RNA viruses with icosahedral, two-layered capsids and protruding fibers. Virus particles are 70 nm in diameter (Fig. 9.3).

Environmental factors: Any ready-to-eat food can be a transmission source if contaminated with virus particles. Virus particles are passed from infected hosts in fecal material.

Type of foodborne illness: Symptoms are similar to those of norovirus infections and include vomiting and diarrhea. Infective doses may be as low as ten virus particles. The incubation time leading to symptom development ranges from 1 to 3 days, and symptoms can last about 1 week. An infected host excretes high concentrations of virus particles in fecal material, and poor hygiene can result in transmission of large numbers of virus particles to surfaces and foods.

Rotavirus is endemic worldwide, and is a leading cause of severe diarrhea among infants and children. Most transmission is from contaminated water, but can also occur through contamination of ready-to-eat foods.

Foodborne illness statistics: Foodborne rotaviruses cause 15,433 illnesses, 348 hospitalizations, and zero deaths per year in the United States (Scallon *et al.*, 2011). Particles typically needed to be concentrated to reach detection limits. Africa and Asia are endemic areas (Nelson and Murphy, 2015).

Costs of foodborne illness: Rotavirus infections were estimated to cost US$1,154/case in the United States in 2010 (Scharff, 2012).

Testing methods:

> The virus has not been isolated from any food associated with an outbreak, and no satisfactory method is available for routine analysis of food. However, it should be possible to apply procedures that have been used to detect the virus in water and in clinical specimens, such as enzyme immunoassays, gene probing, and PCR amplification to food analysis (FDA, 2015a).

Controls: Effective controls to lower the risk of foodborne illness include the following:

- Good hygiene and sanitation practices prevent contamination.
- The virus particles are sensitive to cooking.
- Keep food handlers who are ill away from food preparation areas.

Prions

Microscopic morphology: Prions are proteins that can be found naturally in cell membranes, particularly in neural cells. Prion diseases occur when normal cell prions contact a misfolded form of the protein, which can induce misfolding in the normal prions. The word prion is derived from proteinaceous and infectious, and prions were initially described as proteinaceous, infectious particles (Momcilovic, 2010).

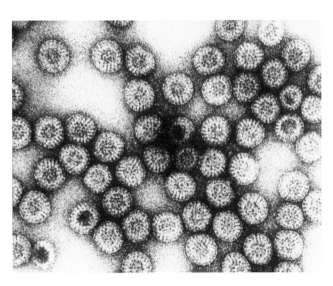

Fig. 9.3. Transmission electron micrograph of rotavirus particles (CDC PHIL, 2014; from Erskine L. Palmer).

Environmental factors: Prion proteins from the remains of infected animals can persist in the environment (Pritzkow *et al.*, 2015).

Type of foodborne illness: Prions cause transmissible spongiform encephalopathies (microscopic holes in brain tissue resembling the texture of sponges), which are deadly neurodegenerative diseases. Prions cause such encephalopathies in sheep and goats in the form of scrapie, as chronic wasting disease in deer, and in cattle in the form of bovine spongiform encephalopathy (BSE; Fig. 9.4). Humans can transmit prion diseases to other humans through cannibalistic practices. There is uncertainty regarding how prions may be passed across species lines (Kong *et al.*, 2008; EFSA, 2015). The crossing of species for BSE prions from cattle to humans has been documented and results in a rare and fatal form known as variant Creutzfeldt–Jakob disease (vCJD). See Box 9.2 for more on the emergence of vCJD in humans.

Infectious dosage is not known. The onset to symptoms can take a decade or more after exposure. Initial symptoms include psychiatric problems followed by neurologic signs including sensory oddities and loss of muscular coordination, followed by forgetfulness. The disease continues to debilitate the victim until death occurs. The duration of symptoms can take up to 2 years.

Foodborne illness statistics: As of 2015, 228 people in the world have been confirmed to have vCJD (FDA, 2015b). It is a very rare disease. The mortality rate is 100%, and there is no cure for the disease. Only three cases of vCJD have been reported in the United States, and initial exposures to prions are suspected to have occurred in other countries for these cases (CDC, 2015b).

Costs of foodborne illness: There is a lack of information in the literature regarding the cost per case for this type of foodborne illness.

Testing methods: According to the United States Food and Drug Administration:

> There are no tests available to determine if food derived from cattle contain the BSE-causing agent. There are postmortem tests to determine if asymptomatic cattle are carrying the BSE disease-causing prions. High-risk tissues for BSE contamination include the cattle's skull, brain, trigeminal ganglia (nerves attached to the brain), eyes, tonsils, spinal cord, dorsal root ganglia (nerves attached to the spinal cord), and the distal ileum (part of the small intestine) (FDA, 2012).

There are immunological-based testing methods available for screening central nervous system tissue in cattle. These are used by the United States Department of Agriculture (USDA, 2012).

Controls: Effective controls to lower the risk of foodborne illness include the following:

- Good agricultural practices prevent food and feed contamination.
- Purchase meat from a reputable supplier.

Fig. 9.4. Photograph of a cow with BSE demonstrating loss of weight, abnormal posture and incoordination (CDC PHIL, 2014; from Art Davis).

Box 9.2. Mad cow disease

In 1986, British cattle showed symptoms similar to those of scrapie in sheep and goats, and the term "mad cow disease" was first used as the colloquialism for BSE. By 1988, over 400 cattle were diagnosed with the disease, and by 1993, over 120,000 were identified with the disease. During the early 1990s, deaths among house cats and zoo animals were linked to beef by-products. In May 1995, there were three human deaths from a newly discovered version of vCJD. By 1996, these and ten other active cases were linked to consuming British beef. This was a terrible blow to the British beef producers, and a 3.5-year international ban on British beef ensued. (CFFS, 2016).

In 2003, the USDA banned sick and injured cattle from the human food supply. Also banned were tissues, such as brain and spinal cord, from cattle over 30 months old and mechanically separated meat. In 2012, the USDA established a rule requiring most livestock traveling across state lines to be tagged for traceability of mad cow disease (CFFS, 2016).

Summary

Viruses and prions are infectious agents that are unable to replicate on their own. Foodborne viruses are extremely contagious and easily transmitted in ready-to-eat foods. The viruses of most concern in food safety include noroviruses, HAV, and rotaviruses. Prions only rarely cause disease in humans, and current good agricultural practices prevent diseases like vCJD.

Further Reading

FDA (2012) *Bad Bug Book*, 2nd edn. United States Food and Drug Administration, Washington, DC. Available at: http://www.fda.gov/Food/FoodborneIllnessContaminants/CausesOfIllnessBadBugBook/default.htm (accessed 29 December 2015).

Juneja, V.K. and Sofos, J.N. (2010) *Pathogens and Toxins in Foods: Challenges and Interventions*. ASM Press, Washington, DC.

References

ASM (2014) Viruses. *Microbe World*. Available at: http://www.microbeworld.org/types-of-microbes/viruses (accessed 29 December 2015).

CDC (2015a) Global burden of norovirus and prospects for vaccine development. Available at: http://www.cdc.gov/norovirus/index.html (accessed 29 December 2015).

CDC (2015b) Variant Creutzfeldt-Jakob Disease (vCJD). Available at: http://www.cdc.gov/prions/vcjd/index.html (accessed 29 December 2015).

CDC PHIL (2014) United States Centers for Disease Control Public Health Image Library. Available at: http://phil.cdc.gov/phil/home.asp (accessed 29 December 2015).

CFFS (2016) Timeline of mad cow disease outbreaks. Available at: http://www.centerforfoodsafety.org/issues/1040/mad-cow-disease/timeline-mad-cow-disease-outbreaks (accessed 30 March 2016).

EFSA (2015) Scientific opinion on a request for a review of a scientific publication concerning the zoonotic potential of ovine scrapie prions: European food safety authority panel on biological Hazards. *EFSA Journal* 13, 4197–4255.

FDA (2015a) BBB – Rotavirus. United States Food and Drug Administration Bad Bug Book. Available at: http://www.fda.gov/Food/FoodborneIllnessContaminants/CausesOfIllnessBadBugBook/ucm071331.htm (accessed 11 October 2016).

FDA (2015b) Revised Preventive Measures to Reduce the Possible Risk of Transmission of Creutzfeldt-Jakob Disease and Variant Creutzfeldt-Jakob Disease by Blood and Blood Products: Guidance for Industry. Available at http://www.fda.gov/downloads/BiologicsBloodVaccines/GuidanceComplianceRegulatoryInformation/Guidances/Blood/UCM307137.pdf (accessed 13 January 2017).

Halfmann, R. (2014) The bright side of prions. *The Scientist* Available at: http://www.the-scientist.com/?articles.view/articleNo/38721/title/The-Bright-Side-of-Prions/ (accessed 29 December 2015).

Hall, A.J. and Lopman, B. (2015) Chapter 3. Infectious diseases related to travel: norovirus. In: *The Yellow Book: CDC Health Information for International Travel 2016*. United States Centers for Disease Control and Prevention, Washington, DC. Available at: http://wwwnc.cdc.gov/travel/page/yellowbook-home-2014 (accessed 29 December 2015).

Kong, Q., Zheng, M., Casalone, C., Qing, L., Huang, S., Chakraborty, B., Wang, P., Chen, F., Cali, I., Corona, C., Martucci, F., Iulini, B., Acutis, P., Wang, L., Liang, J., Wang, M., Li, X., Monaco, S., Zanusso, G., Zou, W.Q., Caramelli, M., and Gambetti, P. (2008) Evaluation of the human transmission risk of an atypical bovine spongiform encephalopathy prion strain. *Journal of Virology* 82, 3697–3701.

Kuritsky, J.N., Oserholm, M.T., Greenberg, H.B., Korlath, J.A., Godes, J.R., Hedberg, C.W., Forfang, J.C., Kapikian, A.Z., McCullough, J.C., and White, K.E. (1984) Norwalk gastroenteritis: a community outbreak associated with bakery product consumption. *Annals of Internal Medicine* 100, 519–521.

Momcilovic, D. (2010) Chapter 22. Prions and prion diseases. In: Juneja, V.K. and Sofos, J.N. (eds) *Pathogens and Toxins in Foods: Challenges and Interventions*. ASM Press, Washington, DC, pp. 343–356.

Nelson, N.P. and Murphy, T.V. (2015) Chapter 3. Infectious diseases related to travel: hepatitis A. In: *The Yellow Book: CDC Health Information for International Travel 2016*. United States Centers for Disease Control and Prevention, Washington, DC. Available at: http://wwwnc.cdc.gov/travel/page/yellowbook-home-2014 (accessed 29 December 2015).

Pritzkow, S., Morales, R., Moda, F., Khan, U., Telling, G.C., Hoover, E., and Soto, C. (2015) Grass plants bind, retain, uptake, and transport infectious prions. *Cell Reports* 11, 1–8.

Richards, G.P., Cliver, D.O., and Greening, G.E. (2015) Chapter 44. Foodborne viruses. In: Salfinger, Y. and Tortorello, M.L. (eds) *Compendium of Methods for the Microbiological Examination of Foods*. Association of Public Health Laboratories, Washington, DC. pp. 595–616.

Scallon, E., Hoekstra, R.M., Angulo, F.J., Tauxe, R.V., Widdowson, M.A., Roy, S.L., Jones, J.L., and Griffin, P.M. (2011) Foodborne illness acquired in the United States – major pathogens. *Emerging Infectious Diseases* 17, 7–15.

Scharff, R.L. (2012) Economic burden from health losses due to foodborne illness in the United States. *Journal of Food Protection* 75, 123–131.

USDA (2012) BSE (Mad Cow Disease). Ongoing Surveillance Information Center. Available at: http://www.usda.gov/wps/portal/usda/usdahome?contentid=BSE_Ongoing_Surveillance_Information_Center.html (accessed 29 December 2015).

Villarreal, L.P. (2008) Are viruses alive? *Scientific American*. Available at: http://www.scientificamerican.com/article/are-viruses-alive-2004/ (accessed 29 December 2015).

WHO (2016) Global alert and response (GAR): Hepatitis A. Available at: http://www.who.int/csr/disease/hepatitis/whocdscsredc2007/en/index2.html (accessed 1 March, 2016).

Woods, J. (2014) Chapter 26B: detection of hepatitis A virus in foods. In: *Bacteriological Analytical Manual (BAM)*. United States Food and Drug Administration, Washington, DC. Available at: http://www.fda.gov/Food/FoodScienceResearch/LaboratoryMethods/ucm2006949.htm (accessed 29 December 2015).

10 Control Measures: The Case of PR/HACCP

Key Questions

- What is HACCP and how has this food safety policy evolved?
- What are the principles of a HACCP plan?
- How was mandatory PR/HACCP policy formulated and implemented?
- What are the cost challenges with PR/HACCP and how were these reduced?
- Is continuous improvement possible with mandatory regulation?

Background on PR/HACCP

We indicated in Chapter 4 that policy is at the center of risk analysis and the interface model. In this chapter, we examine food safety challenges in the pre- and post-pathogen reduction (PR)/hazard analysis and critical control points (HACCP) era. The chapter is organized into three major sections. The first section presents a review of legislative changes in the United States to facilitate effective implementation of PR/HACCP regulation. The next section presents a review of cost–benefit analysis and market impact studies of PR/HACCP regulation. The final section provides a review of studies linking PR/HACCP and public health benefits. The summary section discusses the challenge for continuous improvement of the United States food supply.

What is HACCP?

An old saying:
An ounce of prevention is worth a pound of cure
(Benjamin Franklin).

Nganje (2004) noted that HACCP is an internationally recognized, systematic approach used as a means to assure food safety through the prevention of foodborne hazards. The concept of HACCP was developed in the 1960s through a collaboration between the Pillsbury Company, the United States Army Laboratories and the United States National Aeronautics and Space Administration, where foods for astronauts being developed needed to be as safe as possible (*imagine having a foodborne illness in space*). Low-acid canned foods were the first sector of the commercial food industry to adapt the principles of HACCP in the 1970s with control for the hazard of botulism being the primary goal.

HACCP is about prevention and continuous improvements. How HACCP works is described as:

Significant hazards for a particular food product are identified after a review of all the processing steps and use of scientific information. The steps at which these hazards can be controlled are identified, and critical limits, such as process temperatures and hold times, at key process steps are set. Monitoring procedures are implemented to evaluate conformance with these critical limits. Should the process fall outside these limits, pre-planned corrective actions are taken to prevent the potentially defective product from entering the commerce stream. In addition, the HACCP system relies on extensive verification and documentation to assure that food safety has not been compromised during any step. Thus, HACCP provides a risk assessment structure for putting controls in place to minimize such risks (Goodrich-Schneider *et al.*, 2015).

Evolution of Food Safety Regulations in the Pre- and Post-PR/HACCP Era

Early inspection procedures were through sight and smell. However, an increase in the occurrences of food outbreaks from meat significantly surfaced to the public's attention in the United States with an *Escherichia coli* outbreak (Jack in the Box) in late 1992 to early 1993. The "Jack in the Box" outbreak consisted of 400 cases resulting in more than 100 hospitalizations and the death of one child. It was called Jack in the Box because the cause was traced back to undercooked hamburgers tainted with *E. coli* from a chain of fast food restaurants called Jack in the Box (MacDonald and Osterholm, 1993). This and subsequent outbreaks led to the PR rule and the national implementation of HACCP for the meat industry.

Following the Jack in the Box outbreak and other microbial outbreaks, MacDonald *et al.* (1995) noted that

...in the pre pathogen reduction rule and hazard analysis and critical control points (PR/HACCP) era, the dual goals of public health and consumer protection of meat inspection were centered on hormonal problems, sick animals' inspection with organoleptic methods (using sight, touch and smell), product adulteration with fillers, and deceptive labeling practices... Microbial contamination is the principal

concern today, with pathogens that can cause human illness and are carried in the intestinal tracts of healthy animals. This problem arises because high throughput slaughter and processing methods often combine meat from many carcasses, raising the likelihood of pathogens spread through cross-contamination.

Consequently, inspection with traditional organoleptic methods in the pre-PR/HACCP era were costly and inefficient and were inadequate to prevent contamination of food by pathogens. Process- and performance-based control inspection methods like PR/HACCP may be the most appropriate inspection procedure.

Timeline for the final PR/HACCP rule

In an effort to reduce foodborne microbial pathogens, the National Advisory Committee on Microbiological Criteria for Foods was the first to develop HACCP principles for food production in November 1989 (NACMCF, 1992). On 4 July 1996, the USDA published the final rule for meat and poultry (Federal Register, 1996). As of 25 January 2000, all federal- and state-inspected meat processing plants in the United States operate under mandatory PR/HACCP regulation. See Box 10.1 for the steps leading to implementation of HACCP regulations in the United States.

Box 10.1. United States policy timeline for HACCP

These are the major policy milestones for HACCP in the United States:

- 1970s – HACCP principles were being voluntarily implemented by firms that produced low-acid canned foods.
- 1995 – The Food and Drug Administration (FDA) issued HACCP regulations for fish and seafood.
- 1998 – The United States Department of Agriculture's Food Safety and Inspection Service (USDA-FSIS) mandated HACCP for raw meat and poultry plants. The implementation dates varied by firm size. In large establishments, defined as all establishments with 500 or more employees, implementation was on 26 January 1998. Implementation for smaller establishments, defined as all establishments with ten or more employees but fewer than 500, was on 25 January 1999. In very small establishments, defined as all establishments with fewer than ten

employees or annual sales of less than US$2.5 million, implementation was on 25 January 2000. The Sanitation Standard Operating Procedures (SOP) regulation was applicable on 27 January 1997. The *E. coli* process control testing regulations were applicable on 27 January 1997. The *Salmonella* pathogen reduction performance standard regulations were applicable simultaneously with applicability dates for implementation of HACCP. It should be noted that custom or game animal processing facilities are exempt from mandatory PR/HACCP.
- 2001 – The FDA issued HACCP regulations for juice processing and packaging. Other pilot and voluntary initiatives started in the dairy and other sectors.
- 2010 – The Food Safety Modernization Act will result in more mandatory HACCP-like programs under the FDA and USDA-FSIS.

Nganje (2004) noted that PR/HACCP systems were anticipated to minimize current contamination problems and fulfill the intended inspection purpose of providing a product that is safe when properly handled and prepared for consumption. Under the regulation, processors have the primary responsibility of development and implementation of HACCP systems for meat animal slaughter, carcass fabrication, packaging, and distribution. The major role of the regulatory agency is to verify that a processor's HACCP system is effective and working as intended. The final rule has three major components: (i) a requirement for sanitation standard operating procedures (sanitation SOPs); (ii) development of a HACCP plan with critical control points and critical limits; and (iii) microbiological performance criteria and standard, with a major emphasis on *Salmonella* reduction and the use of *E. coli* testing for process verification.

HACCP Principles and Design

HACCP implementation is based on a good understanding of the seven HACCP principles. Five preliminary steps should be completed before outlining the seven principles of HACCP. These are: (i) develop a HACCP team; (ii) describe the food and its method of distribution; (iii) identify the intended use and target consumers; (iv) develop a flow diagram that describes the process; and (v) verify the flow diagram (NACMCF, 1992) (Appendix 10.1). The seven principles of a HACCP system are:

1. Conduct a hazard analysis. This is where potential hazards are identified as well as effective controls for the identified hazards.
2. Identify the critical control points (CCPs). These are the points at which effective controls can be implemented.
3. Establish critical limits for each CCP. This makes explicit the acceptable limitations of the controls.
4. Establish procedures to monitor CCPs. This verifies and documents that the CCPs are functioning as intended.
5. Establish corrective actions when CCP limits are exceeded. This makes explicit and documents the actions that need to happen to address why limits were exceeded.
6. Establish verification procedures. These are explicitly described procedures that confirm that the CCPs are working as intended.

7. Record keeping. This is all records of the HACCP plan, implementation, progress, and improvement over time.

The detailed procedure and protocol for each principle as discussed by the FDA are presented.

Conduct a hazard analysis (Principle 1)

The FDA indicated that the purpose of the hazard analysis is to develop a list of hazards that are of such significance that they are reasonably likely to cause injury or illness if not effectively controlled. Hazards that are not reasonably likely to occur would not require further consideration within a HACCP plan. It is important to consider in the hazard analysis the ingredients and raw materials, each step in the process, product storage and distribution, and final preparation and use by the consumer. When conducting a hazard analysis, safety concerns must be differentiated from quality concerns. A hazard is defined as a biological, chemical, or physical agent that is reasonably likely to cause illness or injury in the absence of its control. Thus, the word hazard as used in this document is limited to safety.

A thorough hazard analysis is the key to preparing an effective HACCP plan. If the hazard analysis is not done correctly and the hazards warranting control within the HACCP system are not identified, the plan will not be effective regardless of how well it is followed.

The hazard analysis and identification of associated control measures accomplish three objectives. Those hazards and associated control measures are identified. The analysis may identify needed modifications to a process or product so that product safety is further assured or improved. The analysis provides a basis for determining CCPs in Principle 2.

The process of conducting a hazard analysis involves two stages. The first, hazard identification, can be regarded as a brain-storming session. During this stage, the HACCP team reviews the ingredients used in the product, the activities conducted at each step in the process and the equipment used, the final product and its method of storage and distribution, and the intended use and consumers of the product. Based on this review, the team develops a list of potential biological, chemical, or physical hazards that may be introduced, increased, or controlled at each step in the production process. Appendix 10.2 lists examples of questions that may be helpful to consider when identifying potential hazards. Hazard identification focuses on developing a list of potential hazards associated with each process step under direct control of the food operation. A knowledge of any adverse health-related events historically associated with the product will be of value in this exercise.

After the list of potential hazards is assembled, stage two, the hazard evaluation, is conducted. In stage two of the hazard analysis, the HACCP team decides which potential hazards must be addressed in the HACCP plan. During this stage, each potential hazard is evaluated based on the severity of the potential hazard and its likely occurrence. Severity is the seriousness of the consequences of exposure to the hazard. Considerations of severity (e.g. impact of sequelae and magnitude and duration of illness or injury) can be helpful in understanding the public health impact of the hazard. Consideration of the likely occurrence is usually based upon a combination of experience, epidemiological data, and information in the technical literature. When conducting the hazard evaluation, it is helpful to consider the likelihood of exposure and severity of the potential consequences if the hazard is not properly controlled. In addition, consideration should be given to the effects of short-term as well as long-term exposure to the potential hazard. Such considerations do not include common dietary choices that lie outside of HACCP. During the evaluation of each potential hazard, the food, its method of preparation, transportation, storage, and persons likely to consume the product should be considered to determine how each of these factors may influence the likely occurrence and severity of the hazard being controlled. The team must consider the influence of likely procedures for food preparation and storage and whether the intended consumers are susceptible to a potential hazard. However, there may be differences of opinion, even among experts, as to the likely occurrence and severity of a hazard. The HACCP team may have to rely upon the opinion of experts who assist in the development of the HACCP plan.

Hazards identified in one operation or facility may not be significant in another operation producing the same or a similar product. For example, due to differences in equipment and/or an effective maintenance program, the probability of metal contamination may be significant in one facility but not in another. A summary of the HACCP team deliberations and the rationale developed during the hazard analysis should be kept for future reference.

This information will be useful during future reviews and updates of the hazard analysis and the HACCP plan.

Appendix 10.2 is for illustration purposes to further explain the stages of hazard analysis for identifying hazards. Hazard identification and evaluation, as outlined in Appendix 10.3, may eventually be assisted by biological risk assessments as they become available. While the process and output of a risk assessment (NACMCF, 1997) is significantly different from a hazard analysis, the identification of hazards of concern and the hazard evaluation may be facilitated by information from risk assessments. Thus, as risk assessments addressing specific hazards or control factors become available, the HACCP team should take these into consideration.

Upon completion of the hazard analysis, the hazards associated with each step in the production of the food should be listed along with any measure(s) that are used to control the hazard(s). The term control measure is used because not all hazards can be prevented, but virtually all can be controlled. More than one control measure may be required for a specific hazard. On the other hand, more than one hazard may be addressed by a specific control measure (e.g. pasteurization of milk).

For example, if a HACCP team were to conduct a hazard analysis for the production of frozen cooked beef patties (Appendices 10.1 and 10.2), enteric pathogens (e.g. *Salmonella* and verotoxin-producing *E. coli*) in the raw meat would be identified as hazards. Cooking is a control measure that can be used to eliminate these hazards. Table 10.1 is an excerpt from a hazard analysis summary table for this product.

The hazard analysis summary could be presented in several different ways. One format is a table such as the one given in Table 10.1. Another could be a narrative summary of the HACCP team's hazard analysis considerations and a summary table listing only the hazards and associated control measures.

Determine critical control points (Principle 2)

The FDA defines a critical control point as a step at which control can be applied and is essential to prevent or eliminate a food safety hazard or reduce it to an acceptable level. The potential hazards that are reasonably likely to cause illness or injury in the absence of their control must be addressed in determining CCPs.

Table 10.1. Excerpt from a hazard analysis table.

Step	Potential hazard(s)	Justification	Hazard to be addressed in plan? Y/N	Control measure(s)
5. Cooking	Enteric pathogens: e.g. *Salmonella*, verotoxigenic *E. coli*	Enteric pathogens have been associated with outbreaks of foodborne illness from undercooked ground beef	Y	Cooking

Complete and accurate identification of CCPs is fundamental to controlling food safety hazards. The information developed during the hazard analysis is essential for the HACCP team in identifying which steps in the process are CCPs. One strategy to facilitate the identification of each CCP is the use of a CCP decision tree (an example of a decision trees is given in Appendix 10.4). Although application of the CCP decision tree can be useful in determining if a particular step is a CCP for a previously identified hazard, it is merely a tool and not a mandatory element of HACCP. A CCP decision tree should be validated with expert knowledge about risks.

CCPs are located at any step where hazards can be either prevented, eliminated, or reduced to acceptable levels. Examples of CCPs may include thermal processing, chilling, testing ingredients for chemical residues, product formulation control, and testing products for metal contaminants. CCPs must be carefully developed and documented. In addition, they must be used only for purposes of product safety. For example, a specified heat process, at a given time and temperature designed to destroy a specific microbiological pathogen, could be a CCP. Likewise, refrigeration of a pre-cooked food to prevent hazardous microorganisms from multiplying, or the adjustment of a food to a pH necessary to prevent toxin formation could also be CCPs. Different facilities preparing similar food items can differ in the hazards identified and the steps that are CCPs. This can be due to differences in each facility's layout, equipment, selection of ingredients, processes employed, and experience of the HACCP team.

Establish critical limits (Principle 3)

The FDA identifies a critical limit as a maximum and/or minimum value to which a biological, chemical, or physical parameter must be controlled at a CCP to prevent, eliminate, or reduce to an acceptable level the occurrence of a food safety hazard. A critical limit is used to distinguish between safe and unsafe operating conditions at a CCP. Critical limits should not be confused with operational limits that are established for reasons other than food safety.

Each CCP will have one or more control measure to assure that the identified hazards are prevented, eliminated, or reduced to acceptable levels. Each control measure has one or more associated critical limit. Critical limits may be based upon factors such as temperature, time, physical dimensions, humidity, moisture level, water activity, pH, titratable acidity, salt concentration, available chlorine, viscosity, preservatives, or sensory information such as aroma and visual appearance. Critical limits must be scientifically based. For each CCP, there is at least one criterion for food safety that is to be met. An example of a criterion is a specific lethality of a cooking process to reduce *Salmonella*. The critical limits and criteria for food safety may be derived from sources such as regulatory standards and guidelines, literature surveys, experimental results, and experts.

An example is the cooking of beef patties (Appendix 10.5). The process should be designed to ensure the production of a safe product. The hazard analysis for cooked meat patties identified enteric pathogens (e.g. verotoxigenic *E. coli* such as *E. coli* O157:H7 and *Salmonella*) as significant biological hazards. Furthermore, cooking is the step in the process at which control can be applied to reduce the enteric pathogens to an acceptable level. To ensure that an acceptable level is consistently achieved, accurate information is needed on the probable number of the pathogens in the raw patties, their heat resistance, the factors that influence the heating of the patties, and the area of the patty that heats the slowest. Collectively, this information forms the scientific basis for the critical limits that are established. Some of the factors that may affect the thermal destruction of enteric pathogens are listed in Table 10.2. In this example, the HACCP team concluded

Table 10.2. Factors that affect thermal destruction of enteric pathogens.

Process step	CCP	Critical limits
5. Cooking	YES	Oven temperature:___° FTime; rate of heating and cooling (belt speed in ft/min): ____ft/min Patty thickness: ____in. Patty composition: e.g. all beef Oven humidity: ____% RH

that a thermal process equivalent to 155°F (68°C) for 16 seconds would be necessary to assure the safety of this product. To ensure that this time and temperature are attained, the HACCP team for one facility determined that it would be necessary to establish critical limits for the oven temperature and humidity, belt speed (time in oven), and patty thickness and composition (e.g. all beef, beef and other ingredients). Control of these factors enables the facility to produce a wide variety of cooked patties, all of which will be processed to a minimum internal temperature of 155°F (68°C) for 16 seconds. In another facility, the HACCP team may conclude that the best approach is to use the internal patty temperature of 155°F (68°C) and hold for 16 seconds as critical limits. In this second facility, the internal temperature and hold time of the patties are monitored at a frequency to ensure that the critical limits are constantly met as they exit the oven. The example given below applies to the first facility.

Establish monitoring procedures (Principle 4)

Monitoring is a planned sequence of observations or measurements to assess whether a CCP is under control and to produce an accurate record for future use in verification. Monitoring serves three main purposes. First, monitoring is essential to food safety management in that it facilitates tracking of the operation. If monitoring indicates that there is a trend toward loss of control, then action can be taken to bring the process back into control before a deviation from a critical limit occurs. Second, monitoring is used to determine when there is loss of control and a deviation occurs at a CCP (i.e. exceeding or not meeting a critical limit). When a deviation occurs, an appropriate corrective action must be taken. Third, it provides written documentation for use in verification.

An unsafe food may result if a process is not properly controlled and a deviation occurs. Because of the potentially serious consequences of a critical limit deviation, monitoring procedures must be effective. Ideally, monitoring should be continuous, which is possible with many types of physical and chemical methods. For example, the temperature and time for the scheduled thermal process of low-acid canned foods is recorded continuously on temperature recording charts. If the temperature falls below the scheduled temperature or the time is insufficient, as recorded on the chart, the product from the retort is retained and the disposition determined as in Principle 5. Likewise, pH measurement may be performed continually in fluids or by testing each batch before processing. There are many ways to monitor critical limits on a continuous or batch basis and record the data on charts. Continuous monitoring is always preferred when feasible. Monitoring equipment must be carefully calibrated for accuracy.

Assignment of the responsibility for monitoring is an important consideration for each CCP. Specific assignments will depend on the number of CCPs and control measures and the complexity of monitoring. Personnel who monitor CCPs are often associated with production (e.g. line supervisors and/or selected line workers and maintenance personnel) and, as required, quality control personnel. Those individuals must be trained in the monitoring technique for which they are responsible, fully understand the purpose and importance of monitoring, be unbiased in monitoring and reporting, and accurately report the results of monitoring. In addition, employees should be trained in procedures to follow when there is a trend toward loss of control so that adjustments can be made in a timely manner to assure that the process remains under control. The person responsible for monitoring must also immediately report a process or product that does not meet critical limits.

All records and documents associated with CCP monitoring should be dated and signed or initialed by the person doing the monitoring.

When it is not possible to monitor a CCP on a continuous basis, it is necessary to establish a monitoring frequency and procedure that will be reliable enough to indicate that the CCP is under control. Statistically designed data collection or sampling systems lend themselves to this purpose.

Most monitoring procedures need to be rapid because they relate to on-line, "real-time" processes and there will not be time for lengthy analytical testing. Examples of monitoring activities include visual observations and measurement of temperature, time, pH, and moisture level.

Microbiological tests are seldom effective for monitoring due to their time-consuming nature and problems with assuring detection of contaminants. Physical and chemical measurements are often preferred because they are rapid and usually more effective for assuring control of microbiological hazards. For example, the safety of pasteurized milk is based upon measurements of time and temperature of heating rather than testing the heated milk to assure the absence of surviving pathogens.

With certain foods, processes, ingredients, or imports, there may be no alternative to microbiological testing. However, it is important to recognize that a sampling protocol that is adequate to reliably detect low levels of pathogens is seldom possible because of the large number of samples needed. This sampling limitation could result in a false sense of security by those who use an inadequate sampling protocol. In addition, there are technical limitations in many laboratory procedures for detecting and quantitating pathogens and/or their toxins.

Establish corrective actions (Principle 5)

The HACCP system for food safety management is designed to identify health hazards and to establish strategies to prevent, eliminate, or reduce their occurrence. However, ideal circumstances do not always prevail, and deviations from established processes may occur. An important purpose of corrective actions is to prevent foods that may be hazardous from reaching consumers. Where there is a deviation from established critical limits, corrective actions are necessary. Therefore, corrective actions should include the following elements: (i) determine and correct the cause of non-compliance; (ii) determine the disposition of the non-compliant product; and (iii) record the corrective actions that have been taken. Specific corrective actions should be developed in advance for each CCP and included in the HACCP plan. As a minimum, the HACCP plan should specify what is done when a deviation occurs, who is responsible for implementing the corrective actions, and that a record will be developed and maintained of the actions taken. Individuals who have a thorough understanding of the process, product, and HACCP plan should be assigned the responsibility for oversight of corrective actions. As appropriate, experts may be consulted to review the information available

and to assist in determining disposition of the non-compliant product.

Establish verification procedures (Principle 6)

Verification is defined as those activities, other than monitoring, that determine the validity of the HACCP plan and that the system is operating according to the plan. The NAS (1985) pointed out that the major infusion of science in a HACCP system centers on proper identification of the hazards, CCPs, critical limits, and instituting proper verification procedures. These processes should take place during the development and implementation of the HACCP plans and maintenance of the HACCP system. An example of a verification schedule is given in Table 10.3.

One aspect of verification is evaluating whether the facility's HACCP system is functioning according to the HACCP plan. An effective HACCP system requires little end-product testing, since sufficient validated safeguards are built in early in the process. Therefore, rather than relying on end-product testing, firms should rely on frequent reviews of their HACCP plan, verification that the HACCP plan is being correctly followed, and review of CCP monitoring and corrective action records.

Another important aspect of verification is the initial validation of the HACCP plan to determine that the plan is scientifically and technically sound, that all hazards have been identified, and that if the HACCP plan is properly implemented these hazards will be effectively controlled. Information needed to validate the HACCP plan often include: (i) expert advice and scientific studies; and (ii) in-plant observations, measurements, and evaluations. For example, validation of the cooking process for beef patties should include the scientific justification of the heating times and temperatures needed to obtain an appropriate destruction of pathogenic microorganisms (i.e. enteric pathogens) and studies to confirm that the conditions of cooking will deliver the required time and temperature to each beef patty.

Subsequent validations are performed and documented by a HACCP team or an independent expert as needed. For example, validations are conducted when there is an unexplained system failure, a significant product, process, or packaging change occurs, or new hazards are recognized.

In addition, a periodic comprehensive verification of the HACCP system should be conducted by an

Table 10.3. Example of a company established HACCP verification schedule.

Activity	Frequency	Responsibility	Reviewer
Verification activities scheduling	Yearly or upon HACCP system change	HACCP coordinator	Plant manager
Initial validation of HACCP plan	Prior to and during initial implementation of plan	Independent expert(s)[a]	HACCP team
Subsequent validation of HACCP plan	When critical limits changed, significant changes in process, equipment changed, after system failure, etc.	Independent expert(s)[a]	HACCP team
Verification of CCP monitoring as described in the plan (e.g. monitoring of patty cooking temperature)	According to HACCP plan (e.g. once per shift)	According to HACCP plan (e.g. line supervisor)	According to HACCP plan (e.g. quality control)
Review of monitoring, corrective action records to show compliance with the plan	Monthly	Quality assurance	HACCP team
Comprehensive HACCP system verification	Yearly	Independent expert(s)[a]	Plant manager

[a]Done by people other than the team writing and implementing the plan. May require additional technical expertise as well as laboratory and plant test studies.

unbiased, independent authority. Such authorities can be internal or external to the food operation. This should include a technical evaluation of the hazard analysis and each element of the HACCP plan as well as an on-site review of all flow diagrams and appropriate records from operation of the plan. A comprehensive verification is independent of other verification procedures and must be performed to ensure that the HACCP plan is resulting in the control of the hazards. If the results of the comprehensive verification identifies deficiencies, the HACCP team modifies the HACCP plan as necessary.

Verification activities are carried out by individuals within a company, third party experts, and regulatory agencies. It is important that individuals doing verification have appropriate technical expertise to perform this function. The role of regulation and industry in HACCP was further described by the NACMCF (1994).

Examples of verification activities are included as Appendix 10.5.

Establish record keeping and documentation procedures (Principle 7)

The FDA specify that the records maintained for the HACCP system should include the following:

1. A summary of the hazard analysis, including the rationale for determining hazards and control measures.

2. The HACCP plan
Listing of the HACCP team and assigned responsibilities.
Description of the food, its distribution, intended use, and consumer.
Verified flow diagram.
HACCP Plan Summary Table (Table 10.4) that includes information for:
  Steps in the process that are CCPs
  The hazard(s) of concern.
  Critical limits
  Monitoring*
  Corrective actions*
  Verification procedures and schedule*
  Record-keeping procedures*
(* A brief summary of position responsible for performing the activity and the procedures and frequency should be provided.)
3. Support documentation such as validation records.
4. Records that are generated during the operation of the plan.

Examples of HACCP records are given in Appendix 10. 6.

Implementation and maintenance of the HACCP plan

The successful implementation of a HACCP plan is facilitated by commitment from top management.

Table 10.4. Example of a HACCP plan summary table.

CCP	Hazards	Critical limit(s)	Monitoring	Corrective actions	Verification	Records

The next step is to establish a plan that describes the individuals responsible for developing, implementing, and maintaining the HACCP system. Initially, the HACCP coordinator and team are selected and trained as necessary. The team is then responsible for developing the initial plan and coordinating its implementation. Product teams can be appointed to develop HACCP plans for specific products. An important aspect in developing these teams is to assure that they have appropriate training. The workers who will be responsible for monitoring need to be adequately trained. Upon completion of the HACCP plan, operator procedures, forms and procedures for monitoring, and corrective action are developed. Often it is a good idea to develop a timeline for the activities involved in the initial implementation of the HACCP plan. Implementation of the HACCP system involves the continual application of the monitoring, record keeping, corrective action procedures, and other activities as described in the HACCP plan.

Maintaining an effective HACCP system depends largely on regularly scheduled verification activities. The HACCP plan should be updated and revised as needed. An important aspect of maintaining the HACCP system is to assure that all individuals involved are properly trained so they understand their role and can effectively fulfill their responsibilities.

PR/HACCP Policy Formulation and Implementation in the United States

In this section, we discuss the process of making PR/HACCP a mandatory policy in the United States. The emphasis is on pathogen reduction, with *Salmonella* as the single performance standard. *E. coli* and other pathogens are used for process verification. The assumption is that, as firms implement HACCP to reduce *Salmonella* to specified levels, other pathogen levels will reduce. FSIS outline nine policy formulation and implementation procedures. These are:

- *FSIS regulatory and inspection reform plan for PR/HACCP*. FSIS had to reform its existing regulations, policies, and directives to be consistent with HACCP principles and with the Agency's intention to rely more heavily on performance standards. Under PR/HACCP, industry assumes full responsibility for determining CCPs and execution. FSIS only monitors the establishments' compliance with the standards and related requirements under HACCP, and verifies process control, and pathogen reduction and control. In the Federal Register of 29 December 1995 (60 FR 67469), FSIS published an advance notice of proposed rulemaking and additional rulemaking proposals describing the Agency's strategy and changes required to implement PR/HACCP.

- *Changes within FSIS*. The scope of food safety activities by FSIS had to extend beyond slaughter and processing establishments to include new hazard prevention approaches that can occur during transportation and distribution, or retail, restaurant, or food service sales of meat and poultry products. These changes necessitated different resource allocation strategy and organizational structure. Reports prepared by FSIS employees containing analysis and recommendations on these topics were described and made available for public comment in the Federal Register of 12 September 1995 (60 FR 47346).

- *Coordination between food safety agencies to incorporate farm-to-table strategy*. In the preamble to its PR/HACCP proposal, FSIS presented a strategy for the control of food safety hazards throughout the continuum of animal production and slaughter, and the processing, distribution, and sale of meat and poultry products. FSIS had historically focused on the manufacturing of meat and poultry products through its inspection program, but the Agency's public health mandate required that the Agency consider a farm-to-table food safety strategy. FSIS and FDA collaborated in the development of standards governing the safety of potentially hazardous foods, including meat and poultry, eggs, and seafood, during transportation and storage. Proper storage, preparation, and cooking of meat and poultry products are essential to achieving the goal of reducing the risk of foodborne illness to the maximum extent possible.

- *Establishing firm-level performance standards.* Using risk assessment, the agency has designed two CCPs for slaughter, three CCPs for packaging and fresh processing, and one CCP for cooked or smoked processing. These processes constitute the current regulatory PR/HACCP. Most cooked or smoked processing firms operate what they term scientific PR/HACCP with five CCPs. *Salmonella* performance requirements for these processes are presented in Table 10.5.

- *FSIS verification.* FSIS also carries out its verification activities by focusing on an establishment's ongoing compliance with HACCP-related requirements. Verification activities include reviewing all establishment monitoring records for a process, reviewing establishment records for a production lot, direct observation of CCP controls as conducted by establishment employees, collecting samples for FSIS laboratory analysis, or verifying establishment verification activities for a process. The continuous monitoring and verification of production processes and controls sets the stage for further food safety improvements.

- *Process control verification.* Firms that slaughter livestock and poultry have an obligation to control the sanitary dressing process so that contamination with fecal material and other intestinal contents is prevented. FSIS inspectors have access to HACCP records to perform process control verifications.

- *Enforcement and due process.* The nature of the enforcement action taken will vary, depending on the seriousness of the alleged violation. Minor violations of the HACCP requirements may be recorded by agency personnel to determine establishment compliance trends. However, advancement with the Food Safety Modernization Act gives power to the agency to stop inspection and shut down firms with major violations.

- *FSIS enforcement strategy.* The objective of FSIS's enforcement policy with respect to microbial testing is to achieve compliance with the regulations. With respect to *Salmonella*, the agency's goal is to achieve pathogen reduction by ensuring that all slaughter and ground product establishments meet the performance standards established by FSIS. FSIS intends to achieve this goal through an enforcement strategy based on an ongoing testing of all establishments.

- *Reassessment.* The final rule requires that establishments revalidate the HACCP plan whenever significant product, process, deviation, or packaging changes require modification of the plan.

Implementation Issues for Small Firms

The FSIS recognizes that many smaller establishments lack the familiarity with HACCP that exists in many larger establishments. Tables 10.6 to 10.8 show the costs for the proposed rule. These costs could impose a significant burden on small firms. Therefore, FSIS planned an array of assistance activities that facilitated implementation of HACCP in "small" and "very small" establishments. FSIS developed 13 generic HACCP models for the major process categories. The generic models were developed specifically to assist small and very small establishments in preparing their HACCP plans

Table 10.5. *E. coli* testing frequencies and *Salmonella* performance standards.

Species/product	Testing frequencies/number carcasses for *E. coli*	*Salmonella* performance standards (% positive)	Maximum number of positive to achieve standard
Cattle	1 test per 300		
Swine	1 test per 1,000		
Chicken	1 test per 22,000		
Turkey	1 test per 3,000	N/A	N/A
Steers/heifers		1.0	1
Cows/bulls		2.7	2
Ground beef		7.5	5
Fresh pork sausage		N/A	N/A
Broilers		20.0	12
Hogs		8.7	6
Ground turkey		49.9	29
Ground chicken		44.6	26

Source: Federal Register, Vol. 61, No. 144.

Table 10.6. Comparison of costs (US$ million) – proposal to final.

Regulatory component	Proposal	Final
I. Sanitation SOPs	175.9[a]	171.9
II. Time–temperature requirements	45.5	0.0
III. Antimicrobial treatments	51.7	0.0
IV. Micro testing	1,396.3[b]	174.1
V. Compliance with *Salmonella* standards	Not separately estimated[c]	55.5–243.5
VI. Compliance with generic *E. coli* criteria	Not applicable	Not separately estimated
HACCP		
Plan development	35.7	54.8
Annual plan reassessment	0.0	8.9
Record keeping (recording, reviewing, and storing data)	456.4	440.5[d]
Initial training	24.2	22.7[d]
Recurring training	0.0	22.1[e]
VII. Additional overtime	20.9	17.5[d]
Subtotal – industry costs	2,206.6	968.0–1,156.0
VIII. FSIS costs	28.6[f]	56.5
Total	2,235.2	1,024.5–1,212.5

Source: Federal Register, Vol. 61, No. 144.
[a]The preliminary analysis included a higher cost estimate for sanitation SOPs (US$267.8 million) that resulted because of a programming error. The cost estimate of US$175.9 million is based on an effective date of 90 days after publication.
[b]The preliminary analysis was based on the premise that microbial testing would be expanded to cover all meat and poultry processing after the HACCP implementation. The proposed rule only required sampling for carcasses and raw ground product. Thus, the cost estimate of US$1,396.3 million was higher than the actual cost of the proposed sampling requirements.
[c]The preliminary analysis accounted for some of the cost of complying with the new standards under the regulatory components of micro testing, antimicrobial treatments, and time and temperature requirements.
[d]These costs are slightly different from the proposal because of changes in the implementation schedule.
[e]FSIS added the cost for recurring training based on the review of public comments.
[f]Based on current estimates for the cost of training, inspector upgrades and US$0.5 million for annual HACCP verification testing.

Table 10.7. Costs for a typical single-shift processing establishment.

Requirement	Development and implementation costs (US$)	Recurring annual costs (US$)
Sanitation SOPs	190	1,242
HACCP plan development	6,958	0
Annual plan reassessment	0	102
Training	2,514	251
Record keeping	0	6,480
Total	9,662	8,075

Source: Federal Register, Vol. 61, No. 144.

(see Box 10.2). Because each HACCP system is developed by an individual establishment for its particular process and practices, the generic models serve only as illustrations, rather than as prescriptive blueprints for a specific HACCP plan. However, they remove much of the guesswork and reduce the costs associated with developing HACCP plans.

Table 10.8. Costs for a typical single-shift combination establishment.

Requirement	Development and implementation costs (US$)	Recurring annual costs (US$)
Sanitation SOPs	190	1,242
Compliance with *Salmonella* standards	0	800
E. coli sampling	1,043	653
HACCP plan development	6,958	0
Annual plan reassessment	0	102
Training	5,028	503
Record keeping	0	5,434
Total	13,219	8,734

Source: Federal Register, Vol. 61, No. 144.

Nganje (2004) also noted that FSIS conducted HACCP demonstration projects for small and very small establishments during the 2-year period following promulgation of the final rule.

As shown in Table 10.6, the two scenarios developed in the final regulatory impact analysis (FRIA) lead to a range in cost estimates of US$55.5 to US$243.5 million to comply with the new pathogen reduction standards for *Salmonella*. The FRIA recognizes that the performance criteria for generic *E. coli* also creates a set of potential costs for slaughter establishments. A line for these costs is shown in Table 10.6 along with the entry that these costs were not separately quantified. Table 10.7 illustrates the cost for a small, single-shift, processing establishment with two distinct production operations other than raw ground product (overall average estimated at 2.29 operations per establishment).

Table 10.8 illustrates the cost for a small, single-shift, combination (slaughter and further processing) establishment that slaughters cattle or swine, but not both, and has a single further processing operation other than ground product. The cost of meeting the pathogen reduction performance standard assumes that the establishment will use a hot water antimicrobial rinse and have one sample per month analyzed at an outside laboratory

(US$33.35 per sample/US$400 per year). The average number of head slaughtered in a low-volume establishment is approximately 5,000 annually. The annual cost for the rinse is US$400.

The final analysis estimates an average annual cost for HACCP monitoring and recording of US$4,030 for low-volume establishments.

PR/HACCP and Continuous Improvement

In the United States, the Foodborne Diseases Active Surveillance Network (FoodNet) is the principal foodborne disease component of the Centers for Disease Control's Emerging Infections Program. In 1995, FoodNet surveillance began in five states (California, Connecticut, Georgia, Minnesota, and Oregon). Each year, the surveillance area has been expanded by including additional counties or additional states such as New York and Maryland in 1998, Tennessee in 2000, Colorado in 2001, and New Mexico in 2004 (Nganje, 2004). Table 10.9 show infection caused by bacterial pathogens in 2001.

Box 10.2. Provision for small firms

FSIS (2015) made available various HACCP materials to small and very small firms that were used to assist these establishments in conducting their hazard analyses and developing their HACCP plans. These guidance materials include the *Guidebook for the Preparation of HACCP Plans* and *Hazards and Preventive Measures Guide*. These materials were particularly useful to small and very small establishments that may lack the expertise for conducting hazard analyses and designing establishment-specific HACCP plans. The *Guidebook for the Preparation of HACCP Plans* has been designed to provide small and very small establishments with a step-by-step approach for developing a HACCP plan and includes examples and sample forms at each step.

Table 10.9. Infections caused by specific bacterial pathogens by US state.

Pathogen	CA	CO	CT	GA	MD	MN	NY	OR	TN	Total
Campylobacter	999	343	495	614	300	954	248	586	212	4751
Escherichia coli O157	36	37	39	50	16	232	31	77	42	560
Non-O157 STEC	0	4	24	4	0	24	0	5	0	61
Listeria	16	5	15	16	14	4	7	12	5	94
Salmonella	480	317	454	1,675	622	693	271	290	438	5,240
Shigella	427	144	60	714	141	493	28	112	100	2,219
Vibrio	16	5	9	24	13	3	2	5	2	79
Yersinia	17	9	9	50	12	19	6	12	10	144
Total	2,051	878	1,126	3,334	1,147	2,620	608	1,155	836	13,755

STEC, shiga toxin-producing E. coli.
Source: FoodNet Data, 2001.

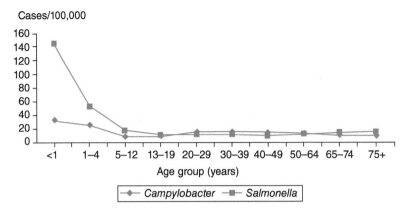

Cases/100,000

Figure 10.1. Incidence of *Campylobacter* and *Salmonella* infections by age group (FoodNet sites, 2001).

FoodNet is used to monitor foodborne diseases and infections caused by bacteria such as *Salmonella, Shigella, Campylobacter, Escherichia coli* O157:H7, *Listeria monocytogenes, Yersinia enterocolitica*, and *Vibrio*, and parasites such as *Cryptosporidium* and *Cyclospora*. The data also provides infection by age group (Fig. 10.1).

The FoodNet data show a high incidence of foodborne diseases in infants and young children, and this constitutes a major concern in public health issues. In 2002, FoodNet began a study that was designed to determine what other opportunities exist for the prevention of foodborne diseases among children, especially for cases of *Salmonella* and *Campylobacter*. Nganje (2004) noted that, in addition to the *E. coli* O157 problem, the increase in the incidence of infections caused by *Salmonella* Newport is a public health challenge. FoodNet is involved with active disease tracking to facilitate continuous improvement of strategies and procedures that could help reduce food safety risks.

procedure used to formulate mandatory HACCP regulation for the meat and poultry industry. PR/HACCP is an improvement over traditional organoleptic methods in the pre-PR/HACCP era. It is a performance-based regulation based on continuous improvement. Provisions have been made in the United States to facilitate PR/HAACP implementation costs for smaller firms by developing generic HACCP models. The major role of the regulatory agency is to verify that a processor's HACCP system is effective and working as intended. Tracking with data from FoodNet suggests that there is a link between PR/HACCP and public health improvements.

References

Federal Register (1996) Pathogen Reduction: Hazard Analysis and Critical Control Points (HACCP) Systems, Final Rule, 25 July, Vols 60, and 61, Nos 23 and 24.

FSIS (2015) Guidebook for the Preparation of HACCP Plans and Generic HACCP Models. Available at: http://www.fsis.usda.gov/wps/portal/fsis/topics/regulatory-compliance/haccp/small-and-very-small-plant-outreach/guidebook-haccp-plans-generic-haccp-models/haccp-plans-guidebook (accessed 30 December 2015).

Goodrich-Schneider, R., Schneider, K.R., and Schmidt, R.H. (2015) HACCP: an overview. Publication #FSHN0512 of the University of Florida Extension

Summary

In this chapter, we have provided information about PR/HACCP, from agencies charged with reducing food safety risks. Information from FDA has provided details on the principles of PR/HACCP. We have also discussed the

Service. Available at: http://edis.ifas.ufl.edu/fs122 (accessed 15 December 2015).

MacDonald K.L. and Osterholm, M.T. (1993) The emergence of *Escherichia coli* O157:H7 infection in the United States: the changing epidemiology of foodborne disease. *Journal of the American Medical Association* 17, 2264–2266.

MacDonald, M.J., Ollinger, E.M., Nelson, K.E., and Handy, R.C. (1995) Structural change in the meat industry: implication for food safety regulations. *Economic Research Service Report.* United States Department of Agriculture, Washington, DC, USA.

NACMCF (1992) Hazard analysis and critical control point system. *International Journal of Food Microbiology* 16, 1–23.

NACMCF (1994) The role of regulatory agencies and industry in HACCP. *International Journal of Food Microbiology* 21, 187–195.

NACMCF (1997) The principles of risk assessment for illness caused by foodborne biological agents. *Journal of Food Protection* 60, 1417–1419.

NAS (1985) *An Evaluation of the Role of Microbiological Criteria for Foods and Food Ingredients.* National Academy of Sciences, National Academy Press, Washington, DC, USA.

Nganje, W.E. (2004) Report on legislative changes, cost-benefit studies and public health improvements of pathogen reduction and hazard analysis and critical control points (PR/HACCP). *Agribusiness and Applied Economics Report.* North Dakota State University, ND, USA.

Appendix 10.1

Examples of common prerequisite programs

The production of safe food products requires that the HACCP system be built upon a solid foundation of prerequisite programs. Each segment of the food industry must provide the conditions necessary to protect food while it is under their control. These conditions and practices are now considered to be prerequisites to the development and implementation of effective HACCP plans. Prerequisite programs provide the basic environmental and operating conditions that are necessary for the production of safe, wholesome food. Common prerequisite programs may include, but are not limited to:

- Facilities: The establishment should be located, constructed, and maintained according to sanitary design principles. There should be linear product flow and traffic control to minimize cross-contamination from raw to cooked materials.

- Supplier control: Each facility should assure that its suppliers have in place effective good manufacturing practice and food safety programs. These may be the subject of continuing supplier guarantee and supplier HACCP system verification.

- Specifications: There should be written specifications for all ingredients, products, and packaging materials.

- Production equipment: All equipment should be constructed and installed according to sanitary design principles. Preventative maintenance and calibration schedules should be established and documented.

- Cleaning and sanitation: All procedures for cleaning and sanitation of the equipment and the facility should be written and followed. A master sanitation schedule should be in place.

- Personal hygiene: All employees and other persons who enter the manufacturing plant should follow the requirements for personal hygiene.

- Training: All employees should receive documented training in personal hygiene, GMP, cleaning and sanitation procedures, personal safety, and their role in the HACCP program.

- Chemical control: Documented procedures must be in place to assure the segregation and proper use of non-food chemicals in the plant. These include cleaning chemicals, fumigants, and pesticides or baits used in or around the plant.

- Receiving, storage, and shipping: All raw materials and products should be stored under sanitary conditions and the proper environmental conditions such as temperature and humidity to assure their safety and wholesomeness.

- Traceability and recall: All raw materials and products should be lot-coded and a recall system in place so that rapid and complete traces and recalls can be done when a product retrieval is necessary.

- Pest control: Effective pest control programs should be in place.

Other examples of prerequisite programs might include quality assurance procedures; SOPs for sanitation, processes, product formulations and recipes; glass control; procedures for receiving, storage, and shipping; labeling; and employee food and ingredient handling practices.

Appendix 10.2

Examples of questions to be considered when conducting a hazard analysis

The hazard analysis consists of asking a series of questions that are appropriate to the process under consideration. The purpose of the questions is to assist in identifying potential hazards.

A. Ingredients

1. Does the food contain any sensitive ingredients that may present microbiological hazards (e.g. *Salmonella*, *Staphylococcus aureus*); chemical hazards (e.g. aflatoxin, antibiotic or pesticide residues); or physical hazards (stones, glass, metal)?
2. Are potable water, ice and steam used in formulating or in handling the food?
3. What are the sources (e.g. geographical region, specific supplier)?

B. Intrinsic factors

Physical characteristics and composition (e.g. pH, type of acidulants, fermentable carbohydrate, water activity, preservatives) of the food during and after processing.

1. What hazards may result if the food composition is not controlled?
2. Does the food permit survival or multiplication of pathogens and/or toxin formation in the food during processing?
3. Will the food permit survival or multiplication of pathogens and/or toxin formation during subsequent steps in the food chain?
4. Are there other similar products in the market place? What has been the safety record for these products? What hazards have been associated with the products?

C. Procedures used for processing

1. Does the process include a controllable processing step that destroys pathogens? If so, which pathogens? Consider both vegetative cells and spores.
2. If the product is subject to recontamination between processing (e.g. cooking, pasteurizing) and packaging, which biological, chemical, or physical hazards are likely to occur?

D. Microbial content of the food

1. What is the normal microbial content of the food?
2. Does the microbial population change during the normal time the food is stored prior to consumption?
3. Does the subsequent change in microbial population alter the safety of the food?
4. Do the answers to the above questions indicate a high likelihood of certain biological hazards?

E. Facility design

1. Does the layout of the facility provide an adequate separation of raw materials from ready-to-eat (RTE) foods if this is important to food safety? If not, what hazards should be considered as possible contaminants of the RTE products?
2. Is positive air pressure maintained in product packaging areas? Is this essential for product safety?
3. Is the traffic pattern for people and moving equipment a significant source of contamination?

F. Equipment design and use

1. Will the equipment provide the time–temperature control that is necessary for safe food?
2. Is the equipment properly sized for the volume of food that will be processed?
3. Can the equipment be sufficiently controlled so that the variation in performance will be within the tolerances required to produce a safe food?
4. Is the equipment reliable or is it prone to frequent breakdowns?
5. Is the equipment designed so that it can be easily cleaned and sanitized?
6. Is there a chance for product contamination with hazardous substances, e.g. glass?
7. What product safety devices are used to enhance consumer safety? E.g.

- metal detectors;
- magnets;
- sifters;
- filters;
- screens;
- thermometers;
- bone removal devices; or
- dud detectors.

8. To what degree will normal equipment wear affect the likely occurrence of a physical hazard (e.g. metal) in the product?

9. Are allergen protocols needed in using equipment for different products?

G. Packaging

1. Does the method of packaging affect the multiplication of microbial pathogens and/or the formation of toxins?

2. Is the package clearly labeled "Keep Refrigerated" if this is required for safety?

3. Does the package include instructions for the safe handling and preparation of the food by the end user?

4. Is the packaging material resistant to damage thereby preventing the entrance of microbial contamination?

5. Are tamper-evident packaging features used?

6. Is each package and case legibly and accurately coded?

7. Does each package contain the proper label?

8. Are potential allergens in the ingredients included in the list of ingredients on the label?

H. Sanitation

1. Can sanitation have an impact upon the safety of the food that is being processed?

2. Can the facility and equipment be easily cleaned and sanitized to permit the safe handling of food?

3. Is it possible to provide sanitary conditions consistently and adequately to assure safe foods?

I. Employee health, hygiene and education

1. Can employee health or personal hygiene practices impact upon the safety of the food being processed?

2. Do the employees understand the process and the factors they must control to assure the preparation of safe foods?

3. Will the employees inform management of a problem that could impact upon safety of food?

J. Conditions of storage between packaging and the end user

1. What is the likelihood that the food will be improperly stored at the wrong temperature?

2. Would an error in improper storage lead to a microbiologically unsafe food?

K. Intended use

1. Will the food be heated by the consumer?

2. Will there likely be leftovers?

L. Intended consumer

1. Is the food intended for the general public?

2. Is the food intended for consumption by a population with increased susceptibility to illness (e.g. infants, the aged, the infirm, immunocompromised individuals)?

3. Is the food to be used for institutional feeding or the home?

Appendix 10.3

Table 10.A1. Examples of how the stages of hazard analysis are used to identify and evaluate hazards.[a]

Hazard analysis stage	Frozen cooked beef patties produced in a manufacturing plant	Product containing eggs prepared for foodservice	Commercial frozen pre-cooked, boned chicken for further processing
Stage 1: Hazard identification			
Determine potential hazards associated with the product	Enteric pathogens (i.e. *E. coli* O157:H7 and *Salmonella*).	*Salmonella* in finished product.	*Staphylococcus aureus* in finished product.
Stage 2: Hazard evaluation			
Assess severity of health consequences if potential hazard is not properly controlled	Epidemiological evidence indicates that these pathogens cause severe health effects including death among children and elderly. Undercooked beef patties have been linked to disease from these pathogens.	Salmonellosis is a foodborne infection causing a moderate to severe illness that can be caused by ingestion of only a few cells of *Salmonella.*	Certain strains of *S. aureus* produce an enterotoxin, which can cause a moderate foodborne illness.
Determine likelihood of occurrence of potential hazard if not properly controlled	*E. coli* O157:H7 is of very low probability and *Salmonella* is of moderate probability in raw meat.	Product is made with liquid eggs that have been associated with past outbreaks of salmonellosis. Recent problems with *Salmonella* serotype Enteritidis in eggs cause increased concern. Probability of *Salmonella* in raw eggs cannot be ruled out. If not effectively controlled, some consumers are likely to be exposed to *Salmonella* from this food.	Product may be contaminated with *S. aureus* due to human handling during boning of cooked chicken. Enterotoxin capable of causing illness will only occur as *S. aureus* multiplies to about 1,000,000 cells per gram. Operating procedures during boning and subsequent freezing prevent growth of *S. aureus*; thus the potential for enterotoxin formation is very low.
Using information above, determine if this potential hazard is to be addressed in the HACCP plan	The HACCP team decides that enteric pathogens are hazards for this product. **Hazards must be addressed in the plan.**	HACCP team determines that if the potential hazard is not properly controlled, consumption of product is likely to result in an unacceptable health risk. **Hazard must be addressed in the plan.**	The HACCP team determines that the potential for enterotoxin formation is very low. However, it is still desirable to keep the initial number of *S. aureus* organisms low. Employee practices that minimize contamination, rapid carbon dioxide freezing, and handling instructions have been adequate to control this potential hazard. **Potential hazard does not need to be addressed in plan.**

[a]For illustrative purposes only. The potential hazards identified may not be the only hazards associated with the products listed. The responses may be different for different establishments.

Appendix 10.4

Example of a CCP decision tree

Important considerations when using the decision tree:

- The decision tree is used after the hazard analysis.
- The decision tree then is used at the steps where a hazard that must be addressed in the HACCP plan has been identified.

- A subsequent step in the process may be more effective for controlling a hazard and may be the preferred CCP.
- More than one step in a process may be involved in controlling a hazard.
- More than one hazard may be controlled by a specific control measure.

Q 1.   Does this step involve a hazard of sufficient likelihood of occurence and severity to warrant its control?

YES   NO → Not a CCP

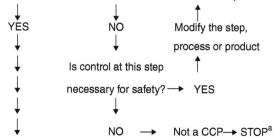

Q 2.   Does a control measure for the hazard exist at this step?

YES   NO   Modify the step,
                                        process or product

                    Is control at this step

                    necessary for safety? → YES

                            NO   →   Not a CCP → STOPᵃ

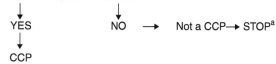

Q 3.   Is control at this step necessary to prevent, eliminate, or reduce the risk of the hazard to consumers?

YES   NO   →   Not a CCP → STOPᵃ

CCP

ᵃ Proceed to next step in the process.

Appendix 10.5

Examples of verification activities

A. Verification procedures may include:

1. Establishment of appropriate verification schedules.
2. Review of the HACCP plan for completeness.
3. Confirmation of the accuracy of the flow diagram.
4. Review of the HACCP system to determine if the facility is operating according to the HACCP plan.
5. Review of CCP monitoring records.

6. Review of records for deviations and corrective actions.
7. Validation of critical limits to confirm that they are adequate to control significant hazards.
8. Validation of HACCP plan, including on-site review.
9. Review of modifications of the HACCP plan.
10. Sampling and testing to verify CCPs.

B. Verification should be conducted:

1. Routinely, or on an unannounced basis, to assure CCPs are under control.

2. When there are emerging concerns about the safety of the product.

3. When foods have been implicated as a vehicle of foodborne disease.

4. To confirm that changes have been implemented correctly after a HACCP plan has been modified.

5. To assess whether a HACCP plan should be modified due to a change in the process, equipment, ingredients, etc.

C. Verification reports may include information on the presence and adequacy of:

1. The HACCP plan and the person(s) responsible for administering and updating the HACCP plan.

2. The records associated with CCP monitoring.

3. Direct recording of monitoring data of the CCP while in operation.

4. Certification that monitoring equipment is properly calibrated and in working order.

5. Corrective actions for deviations.

6. Sampling and testing methods used to verify that CCPs are under control.

7. Modifications to the HACCP plan.

8. Training and knowledge of individuals responsible for monitoring CCPs.

9. Validation activities.

Examples of HACCP records

A. Ingredients for which critical limits have been established.

1. Supplier certification records documenting compliance of an ingredient with a critical limit.

2. Processor audit records verifying supplier compliance.

3. Storage records (e.g. time, temperature) for when ingredient storage is a CCP.

B. Processing, storage, and distribution records

1. Information that establishes the efficacy of a CCP to maintain product safety.

2. Data establishing the safe shelf life of the product, if age of product can affect safety.

3. Records indicating compliance with critical limits when packaging materials, labeling, or sealing specifications are necessary for food safety.

4. Monitoring records.

5. Verification records.

C. Deviation and corrective action records.

D. Employee training records that are pertinent to CCPs and the HACCP plan.

E. Documentation of the adequacy of the HACCP plan from a knowledgeable HACCP expert.

11 Cost of Microbial Foodborne Outbreaks

Key Questions

- Why do we need to estimate the cost of foodborne illness outbreaks and cost-effectiveness for firms to implement control measures?
- How do we quantify the economic cost of a foodborne illness outbreak to a firm?
- How do firm investment costs compare to benefits of reduced illness and deaths?

Introduction to Understanding Costs and Why?

Each year in the United States, there are approximately 48 million illnesses (one in six Americans), 128,000 hospitalizations, and 3,000 deaths due to foodborne illness (CDC, 2011). These losses introduce significant amounts of cost to consumers, firms, and society as a whole. The cost to individuals for each pathogen was presented in Chapters 5 to 9. In this chapter, we extend our discussion to cover the cost of foodborne illness to firms and society as a whole.

Food safety is a pressing issue for governments, food manufacturers, retailers, and consumers as they strive to estimate the costs of food safety recalls to society and to individual firms. From an alternative viewpoint, food safety costs could be viewed as economic benefits of mitigating foodborne illness outbreaks.

So why do regulatory agencies need to estimate the costs and cost-effectiveness of mitigation strategies? Executive Orders or Acts exist that protect firms against costly regulatory procedures:

Executive Order 12291 compels agencies to use cost–benefit analysis as a component of decision making.
The Regulatory Flexibility Act (P.L. 96-354) requires regulatory relief for small businesses where feasible.

The United States Food and Drug Administration (FDA) finds that implementing control measures to reduce foodborne illness costs constitutes a major rule under both the Executive Order and the Regulatory Flexibility Act. The implication is that firms may respond adversely if case-control measures are too costly and threaten their existence in business. Cost-mitigation measures must meet the cost-effectiveness or cost–benefit criteria.

Firm-Level Costs of Foodborne Illness Outbreaks

Firm-level costs are correlated with the type of recall. There are three classes of recall in the United States:

- Class I recalls are those that will cause serious adverse health consequences and deaths.
- Class II recalls are those that cause temporary or medically reversible health effects.
- Class III recalls are likely to not cause any health problems. These are usually product quality issues with mislabeling.

The federal agency guidelines for implementing a recall include: (i) a recall policy; (ii) evaluation of class for the recall; (iii) adoption of a recall strategy; and (iv) explaining the role of the firm(s) involved and discussing guidelines to terminate the recall.

Industry guidelines for implementing a recall requires firms to have: (i) a written recall plan; (ii) a product code system that can trace all products to facilitate a recall; and (iii) records of all products distributed and sold.

Cost concepts

Depreciation

It is important to understand the economic concepts of depreciation when estimating costs (see Box 11.1). When an outcome or cost extends beyond the reporting period (e.g. annually), it must be depreciated. For example, depreciation may represent the annual portion of the fixed cost associated with machinery purchased in 2015 that has an expected life of 20 years.

We introduce three methods for calculating depreciation (or amortizing fixed costs):

- The straight line method of depreciation can be calculated from the relationship:

$$\text{Depreciation} = (OC - SV)/N$$

where OC is the original cost of an item, SV is the salvage or junk value at the end of the lifespan of the item, and N is the lifespan of the item.

- The double declining balance method can be computed from the relationship:

$$\text{Depreciation} = (2/N)^*(R)$$

where R is the remaining book value of the asset at the beginning of the period (e.g. year).

- The sum of year digit method is computed as:

$$\text{Depreciation} = \{(N - t + 1)/[N^*(N+1)/2]\}^* \\ (OC - SV)$$

where t is the current period (e.g. depreciation for the first year is calculated using $t = 1$).

Fixed (implementation) costs

These are costs that do not vary with production levels. Generally, these are expenses on equipment or material that have a life span greater than 1 year. Examples include the costs associated with buildings, machinery, and computers. Depreciation represents fixed costs.

Operating (variable) costs

These are costs that vary with the level of production. They are generally considered to be expenses on equipment or material that has a life span equal to or less than 1 year. These expenses would not be incurred if the business ceased to exist, and increase with production (although not necessarily linearly). Examples include electricity, paper wraps, boxes, and labor.

Cost structure for firms

The costs of foodborne pathogens for firms are much broader than individual costs incurred from illness or death presented in Chapters 6–9. In this section, we discuss the quality cost index (QCI), a common cost structure used by firms to measure investments

---

Box 11.1. Practice problem

A new piece of laboratory equipment costs US$12,000 and has a life span of 4 years. The salvage value of the equipment is US$1,500. Using each of the three depreciation methods, calculate the annual depreciation of the new equipment, and answer the questions below:

1. Which method of depreciation most accelerates expenses?
2. What is the tax advantage of accelerating expenses?

Are your answers the same as below?

Answers

1. The double declining method.
2. Net present value of an expense now is "more valuable" than that of an expense at a later time. A reduction in taxes today is worth more than a reduction in taxes in a future period.

	Straight line method	Double declining balance method	Sum of year digit method
Year 1	US$2,625	US$6,000	US$4,200
Year 2	US$2,625	US$3,000	US$3,150
Year 3	US$2,625	US$1,500	US$2,100
Year 4	US$2,625	US$750	US$1,050

or costs of foodborne risks to a firm and compare these costs to the performance of the firm:

$$\text{Quality cost index} = (\text{total quality costs}/\text{net sales}) \times 100$$

QCI is defined as the ratio of total costs to net sales (Hosking, 1984; Zugarramurdi *et al.*, 1995; Lupin *et al.*, 2010). It is a function of three major cost components (prevention–appraisal–failure; PAF) weighted by net sales. Firms can use the QCI to track the cost-effectiveness of their food safety operations. The method can also be used to compare the total costs and the benefits of food safety investments over time, by examining the QCI trend. Total cost is the sum of PAF costs.

Prevention costs are generally associated with fixed cost items. These could include hazard analysis critical control point (HACCP; see Chapter 10) and hygiene plan development costs, training costs, material, building, and equipment costs. These costs are usually planned and incurred before the actual operation of a food safety system (Lupin *et al.*, 2010).

Appraisal costs are mostly attributed to operating costs. These are costs associated with testing and tracking pathogen levels, implementing critical control limits, and evaluating incoming raw material processes, intermediate and final products, and services to ensure they comply with HACCP plans. They can also include operating hygiene protocols and maintaining the firm's HACCP system to ensure continuous improvement.

Failure costs can be divided into internal and external failure costs. Internal failure costs are costs that indicate inadequate compliance to pathogen control protocols like HACCP plans. Usually these are additional costs from product re-working or retesting. External failure costs are those related to inadequate compliance that could result in type I, II, or III recalls. These costs could include illness or death, litigations, and costs of product recall. Product recall costs include the costs of damaged goods, logistics costs, tracking costs, costs of lost market share and business, and the cost of negative publicity (see the example in Box 11.2).

The failure costs due to a 2006 *E. coli* O157:H7 outbreak (CDC, 2006) were startling. In this case, 199 persons were infected, including 31 cases of hemolytic–uremic syndrome (a serious complication that can cause kidney failure) and three deaths.

The personal injury and lives lost continue to be the atmost regrettable consequence from this outbreak, yet the economic and financial fallout also severely impacted the public. The entire United States spinach industry experienced financial losses and reputation damage, while consumers' confidence in food safety diminished and public funding was allotted to recoup the costs of the outbreak.

The total volume of contaminated product that caused the outbreak and the product recalls was approximately 15,750 lbs of bagged Dole baby spinach, which was identified by the code P227A. The code indicated that the spinach was produced at Natural Selection's south plant (P) on the 227th day of the year (15 August) during the first of two shifts (A) (Weise and Schmit, 2007). Although several leafy green products were recalled due to the outbreak, the 15,750 lbs of bagged Dole baby spinach was deemed responsible for the outbreak and product recalls.

From this single outbreak, we estimate the failure costs associated with the 2006 *E. coli* O157:H7 outbreak based on the United States spinach industry losses and the total costs associated with the product recall to be approximately US$129 million. The industry cost of the outbreak was estimated at US$80 million, but some studies have reported industry losses in the range of US$100 million to over US$350 million (McKinley, 2006; Weise and Schmit, 2007). Annual failure costs for multiple outbreaks in the United States could be in the billions of dollars.

Comparing Firm Costs and Benefits of Reduced Illness and Deaths

We use the case of the meat industry's compliance with regulatory requirements for pathogen reduction (PR)/HACCP (see Chapter 10) for this comparison and introduce other methods and techniques used in empirical analysis to compare cost and benefits when factors are stochastic (i.e. a pattern that may be analyzed statistically but may not be predicted precisely). See Box 11.3 for an example. A stochastic factor could also be referred to as a random variable. Probability functions are used to measure stochastic outcomes. For example, it might be more accurate to provide a range of costs rather than a single number.

The first method described in this section is the MICMAC (microscopic–macroscopic) scenario method, to determine factors that will facilitate

Box 11.2. Estimated failure costs linked to the 2006 *Escherichia coli* O157:H7 outbreak and recall

Recall-related costs	
Retail value of Dole Baby Spinach (per unit of 3 lbs)	US$3.89
Approximate number of units recalled	42,000
TOTAL RECALL-RELATED COSTS	*US$163,380*
Lost productivity expenses	
Lost productivity due to *E. coli* O157:H7 (per case)	US$1,871.96
Approximate number of *E. coli* O157:H7 cases linked to outbreak	204
TOTAL LOST PRODUCTIVITY EXPENSES	*US$381,879.84*
Medical and loss-of-life calculations	
Did not visit physician and survived (per case)	US$28
Estimated unreported cases	6,000
TOTAL	US$168,000
Visited physician and survived (per case)	US$495
Approximate number of cases	100
TOTAL	US$49,500
Did not have HUS and survived (per case)	US$6,550
Approximate number of cases	70
TOTAL	US$458,500
Had HUS and survived (per case)	US$36,525
Approximate number of cases	31
TOTAL	US$1,132,275
Had HUS and did not survive (per case)	US$6,766,498
Approximate number of cases	3
TOTAL	US$20,299,494
TOTAL MEDICAL AND LOSS-OF-LIFE COST	US$22,107,769
Industry lost sales following outbreak and recall	US$80,000,000
Federal funding (within Iraq Bill) to compensate "innocent" farmers	US$25,000,000
USDA grant funding to identify source of outbreak	US$1,200,000
Total estimated failure costs (2006 *E. coli* outbreak)	**US$128,853,028.84**
Approximate vol. of contaminated product (lbs)	15,750

HUS, hemolytic–uremic syndrome; USDA, United States Department of Agriculture.
Source: Compiled from CIDRAP (2008), McKinley (2006).

Box 11.3. Control measures

Currently, food service and retail facilities are implementing various forms of pathogen reduction interventions including voluntary PR/HACCP. These interventions fall into three broad strategy categories: (i) USDA–FDA verification; (ii) contracting with an external firm to perform food safety audits; and (iii) voluntary PR/HACCP. It should be noted that all strategies require standard operating procedures for hygiene. The first strategy requires the USDA–FDA to carry out the random checks and pathogen testing; the second strategy involves engaging an external firm (e.g. Fresh Check) to carry out the random checks and pathogen testing; and the third strategy requires the establishment to have a functional PR/HACCP plan with critical control points and pathogen testing. Given the public concerns on overall food safety, the question raised is whether mandatory regulation at the retail level or voluntary standards would be cost-effective and efficient in reducing food safety risks.

cost-effective food risk reduction strategies. Then we compare the cost-effectiveness of three strategies used by firms with stochastic dominance, an empirical method used in economic analysis.

Cost-effectiveness is defined here as the optimal point at which additional expenditure to reduce pathogen prevalence will have minimal food risk reduction impacts.

MICMAC scenario methods

Scenario methods are used to determine driver and dependent variables (or core factors) that could facilitate an adoption of PR/HACCP at retail facilities or food safety risk reduction strategies. The MICMAC scenario method is used to help determine which variables will have a strong influence and strong dependence on the perceived outcome. Variables with strong influence are also called determinant variables. A determinant or influence variable has a high degree of significance or a large direct impact on the outcome, without being dependent on other factors. A dependent variable, on the other hand, can affect the outcome but it is very sensitive to second- and third-order interaction with the influence variables (Godet, 1993).

This method requires that each variable be assigned an initial weight that corresponds to their impact on other variables. The weights can range from 0 to 3. A weight of 0 means that the variable has no impact on the corresponding variable and 3 means it has a strong impact on another variable. The weights of the variables are determined from the literature or a survey. Weights are put into a matrix and the matrix is multiplied until it is stable. MICMAC then generates a report of the strong driver and dependent variables. Table 11.1 shows sample factors that could be used in PR/HACCP analysis.

Variables that are both strong dependents and have strong influence are the factors that will have the largest impact on implementing a desired strategy like PR/HACCP. Figure 11.1 shows the variables that will have a direct influence on the adoption of PR/HACCP by retail firms. The variables in the upper right quadrant have strong dependence and strong influence on adoption of PR/HACCP. This example shows that the number of foodborne illness outbreaks nationally and the number of retail firms leading the adoption of PR/HACCP plans are the main determinants.

Four variables were found to have an indirect influence and dependence on PR/HACCP implementation (Fig. 11.2): (i) the number of foodborne illness outbreaks nationally; (ii) the number of retail firms adopting HACCP plans; (iii) the public demand for change in food safety policy; and (iv) the probability of a foodborne illness outbreak for an individual firm. This is a very important first step when conducting a cost-effective empirical analysis. The identified variables could not be used in an in-depth empirical analysis.

What food safety risk reduction strategy is cost-effective for retail firms?

A quality loss framework (PAF) is used to quantify the cost for each strategy. For example, supply and demand impacts and the cost of intervention (testing and sampling costs) are quantified. Quality loss costs cover expenditures associated with ensuring that products conform to quality specifications. The conformance costs include costs of prevention and appraisal, while the non-conformance costs include the costs of internal and external failure. Internal failure occurs when *Salmonella* levels (the only performance standard pathogen with PR/HACCP) are higher than the performance standard (e.g. 49% prevalence for ground turkey, 0% prevalence for ready-to-eat meats). The quality loss function is consistent with other quality functions that have been used extensively in the quality management literature (Taguchi, 1986). The loss function is a financial measure of user dissatisfaction with a product's performance as it deviates from a target safety value.

Once the quality loss function is determined for each strategy, a stochastic dominance method is used to rank alternative strategies. The advantage of stochastic dominance is that it incorporates the firm's preference for alternative strategies by using risk aversion coefficients. There are several different types of stochastic dominance, but in this study, stochastic dominance with respect to a function was used. First-degree stochastic dominance uses the mean or average to compare risky outcomes. Second-degree stochastic dominance uses the mean and standard deviation and is comparable to a mean-variance efficient set (dominant set). Third-degree stochastic dominance with respect to a

Table 11.1. Abbreviations for variables affecting the effectiveness of PR/HACCP at the retail level.

Variable	Abbreviation
Cost of labor to train employees	LabCos
Cost of restructuring operations	Restruc
Cost of microbial testing	CosMT
Firm's probability of having a foodborne illness outbreak	ProbOut
The number of firms adopting HACCP plans	#adopt
The price of meat products	Price
The demand for meat products	DemMeat
The supply of meat products	SupMeat
The number of foodborne illness outbreak occurring nationally	#OutNation
The public demand for food policy changes	PublicDem
Consumer education about safe food practices	ConsEd
Diet trends in the United States	Diet

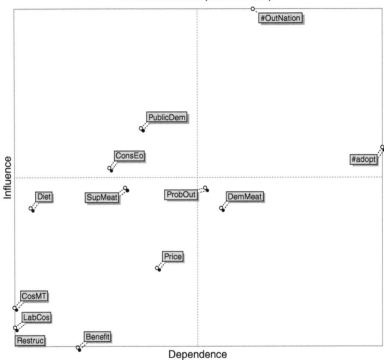

Fig. 11.1. Direct influence/dependence map for retail PR/HACCP implementation. See Table 11.1 for descriptions of abbreviations.

function uses mean returns, variance, and preferences to compare or rank risky outcomes.

When the appropriate rankings of the strategies are found, they will either validate or discredit the hypothesis of PR/HACCP being the most cost-effective mitigation strategy at the retail level.

Quality loss

A quality loss function is used to estimate quality loss due to violations of performance standards. Quality loss could occur at any point along the processing, retailing, and consumption continuum.

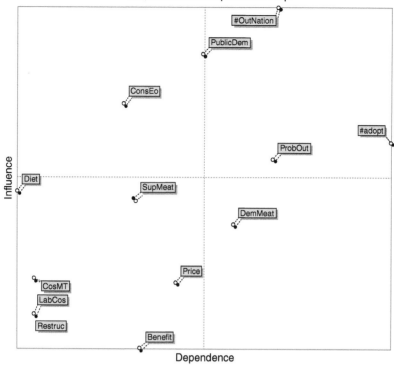

Fig. 11.2. Indirect influence/dependence map for retail PR/HACCP implementation. See Table 11.1 for descriptions of abbreviations.

A Taguchi loss function with smaller-is-better characteristics is used to calculate quality loss. The Taguchi loss function establishes a financial measure of user dissatisfaction with a product's performance as that performance deviates from a target tolerance level. The loss function is defined as $L = (A_o/\Delta_o^2)\sigma^2$, where L is the quality loss, A_o is the welfare loss when the tolerance limit is violated, Δ_o is the tolerance limit, and σ^2 measures the variance of the quality of the product. In these smaller-is-better models, variance is sometimes measured as deviation from the target. Because the data were generated based on a binomial distribution (pathogen present or absent), variance was calculated based on the formula for binomial distributions. The loss to society is composed of costs incurred by the retail firm and the customer. The firm is exposed to rejection costs, loss of future business, etc., while the consumer is exposed to foodborne illness and death. Quality deviations from the target value of zero represent an implicit cost to the system; thus, shipments containing even minimal microbial pathogen content can incur quality loss.

The welfare loss when the tolerance limit is violated is comprised of three major components. The first component is the loss imparted by decreasing demand when an outbreak occurs. Empirical evidence from Kay (2003) showed that decreasing demand is the most important component of the loss because it represents about 60% of the total loss that a firm can incur. The second component is the loss due to a decrease in market price. Studies have found a positive relationship between consumer perceptions about product quality and product price (Kerin et al., 1992; Grewal et al., 2003). Thus, an outbreak causing perceptions of poor food safety quality could substantially lower the prices of the affected products. This price decrease represents about 4.2% of the total cost when there is an outbreak (Kay, 2003). The last component is the cost of recall. Overall, product profitability may be impacted by the overall consumer production evaluation spanning nutrition, food safety, and a host of other variables. A positive image about the food product will instead enhance profitability via increased demand (Burton *et al.*, 2009). When there is an outbreak, the firm may recall all of that day's shipment, estimated as the total revenue (TR) for that day.

The welfare loss, A_o is an additive function of D, the recall impact on consumer demand, P_m, the impact on meat market price, and TR. The TR components of total output and price were modeled as stochastic variables. Total output was based on data collected from local retail meat shops and was modeled as a risk-triangular distribution with values of \$156,250 for a high value, \$75,000 for a low, and \$98,125 for the most likely value. Price was simulated by taking the average monthly prices on each meat type for the years 1970 to 2006 and fitting those numbers to log-normal distributions. Each meat type had a different distribution. In the model, it is assumed that if a test is made with a sampling intensity of at least two samples (the minimum number of samples required to be taken per CCP), the potential quality loss is reduced by 50%. This derives from one important model assumption that when the probability of contamination exceeds zero, there is a 50% reduction in quality loss if appropriate minimal testing and intervention are performed. This, in effect, is a cornerstone assumption in this study that HACCP is at least 50% effective in reducing pathogen levels. FoodNet data reveal that pathogen levels have declined significantly after PR/HACCP implementation. Assumptions on the effectiveness of HACCP have been reported by Marler (2010), Antle (2000), and Knutson *et al.* (1995). Antle (2000) simulated safety levels ranging from 50% to 90%. In addition, the FSIS assumed 10–100% effectiveness for HACCP as a basis for its regulatory impact assessment.

Value of risk reduction for alternative strategy

When microbial testing is done by an agency and the probability of contamination is greater than zero, benefits result from risk reduction. The value of risk reduction could be greater than the TR because recall costs include total value of shipment, loss in market share value, and liability payments. The value of risk reduction is the additional benefit accruing to a firm that tests for pathogens and implements specific intervention strategy. It is a measure of the benefit derived from not shutting down due to an outbreak of a particular pathogen. Hence the value of risk reduction is a function of the testing decision and the sampling intensity, and it estimates the portion of TR that is retained when an outbreak is prevented. It is mathematically defined as:

$$\pi_i = \theta_i * (TR) * \beta_i$$

where π is the value of risk reduction and β is an element of the set $\{0,1\}$, which is a binary testing decision variable, where 1 equals the optimal decision is to test for pathogens and 0 otherwise.

Testing costs

Testing for pathogens occurs at various times randomly with either strategy. Testing may be done at different intensity levels (number of samples) and/or different tolerance levels (number of pathogens at which the product is still considered safe for human consumption). These costs were measured for each strategy. Conventional wisdom is that higher sampling intensities and testing decreases the probability of producing and selling contaminated food products.

Testing costs include three major components: (i) the utilities cost for each strategy; (ii) associated labor costs; and/or (iii) the cost of pathogen testing in laboratories outside the retail firm. Survey findings reveal that, on average, labor costs for different types of pathogen testing were US\$14 per test. However, labor costs can vary between US\$8 and US\$20 per test. Hence labor costs were represented as a risk-triangular distribution in the model, especially because USDA–FDA inspection agents may require more testing if food safety problems persist. The cost of utilities for each strategy is assumed fixed at US\$36 per test. The cost of *Salmonella* and *Campylobacter* testing can vary with the type of test used, ranging between US\$10 and US\$14 per test. Like labor costs, the cost of *Salmonella* and *Campylobacter* testing is also represented by a Risk-Uniform distribution.

The cost of *E. coli* testing can vary from US\$100 to US\$200 per test, depending on the type of test, with the average price being US\$150. Like labor costs, the prevalence of *Salmonella*, *Campylobacter*, and *E. coli* were represented by stochastic variables. *E. coli* tests were assumed to be a risk-triangular distribution with US\$100 being the lowest cost, US\$200 being the highest possible cost, and US\$150 being the most likely testing cost. Total testing costs, C, for each pathogen type are estimated by using the following equation:

$$C_i = (L_i + U_i + T_i)*n_i*\beta_i$$

where L is the labor cost for collecting and preparing product samples, U is the utilities cost, T is the cost of pathogen testing, and n, i, and β are as previously defined.

Total economic costs

The total economic costs associated with the retail meat sector are composed of the value of risk reduction, testing and sampling costs, and the quality loss. The direct cost components include the testing costs, and utilities and labor costs. The indirect cost component accounts for quality loss incurred when there is a violation of the tolerance level. The value of risk reduction was considered a benefit in this study, because it is the cost avoided when adequate testing for pathogens is performed and an intervention strategy is implemented. Hence the total system cost, TC, is defined as:

$$TC = \Sigma L_i + C_i - \pi_i$$

A net benefit function can be developed by subtracting costs and the fixed costs of alternative strategies from the TR from the particular product. Hence the net benefit (NB) function is:

$$NB(\beta, n) = p * Y - TC(\beta, n)$$

The net benefit is simulated for each strategy with data from retail establishments.

Stochastic dominance comparison of net benefits

Results for the three pathogen contamination probabilities and prevalence were generated assuming tolerance levels of 29%, 15%, 10%, 5%, and 1%. Table 11.2. reveals that quality loss estimates increased with tightening of tolerance levels. The highest quality loss values were found at the 1% tolerance level. This is an indication that stricter mandatory compliance could lead to increased quality loss and costs to retailing. On the contrary, higher tolerance levels decrease with the quality loss associated with *E. coli* in all meat types.

The estimates of the value of risk reduction for *E. coli* (for the average retail facility) were by far the highest in beef with values of US$199.97/day at the 29% level, US$233.72/day at the 15% level, US$246.91/day at the 10% level, US$260.77/day at the 5% level, and US$272.37/day at the 1% level. This is because the prevalence of *E. coli* was highest in beef. The next highest estimates came from testing for *E. coli* in pork. This also follows suit because pork had the next highest *E. coli* prevalence. The values for pork were US$84.74 at a 29% tolerance level, US$113.18 at a 15% tolerance level, US$124.38 at a 10% tolerance level, US$136.25 at a 5% tolerance level, and US$146.27 at a 1% tolerance level. *E. coli* testing for chicken had very similar risk reduction values. Typically, the value of risk reduction for chicken was US$3–4 less than the value of risk reduction for pork at each tolerance level. *E. coli* testing for turkey had the lowest value of risk reduction. The values were US$50.90 at a 29% level, US$75.89 at a 15% tolerance level, US$85.96 at a 10% tolerance level, US$96.76 at a 5% tolerance level, and US$105.98 at a 1% tolerance level. Overall, the value of risk reduction from the three intervention strategies was lower than quality loss estimates when compared on a monthly basis. This is a good indication that it is beneficial for retail firms to implement a food safety risk-reducing strategy.

In the case of *Campylobacter* testing, it was only beef that provided a risk reduction of US$19.50 at the 1% tolerance level. For chicken and *Campylobacter*, the values of risk reduction were US$15.90 at a 15% level, US$26.00 at a 10% tolerance level, US$38.75 at a 5% tolerance level and US$51.27 at a 1% tolerance level. *Campylobacter* testing for pork provided a risk reduction only at a 1% tolerance level with a value of US$2.52. Turkey showed risk reduction from *Campylobacter* testing at 5% and 1% tolerance levels with values of US$0.25 and US$0.91, respectively. In the case of *Salmonella* testing, only chicken provided a risk reduction of US$10.73 at a 1% tolerance level. Turkey showed risk reduction from *Campylobacter* testing at 5% and 1% levels with values of US$0.16 and US$0.79, respectively.

Optimal cost-effective intervention strategies at retail with stochastic dominance analysis

The three strategies were compared using stochastic dominance methodologies for alternative meat types and pathogens. These alternatives were compared using SIMETAR software. Upper and lower risk aversion coefficients were used (Pratt, 1964). A lower risk aversion coefficient (RAC) of 0.000001 and an upper RAC of 0.1 were used to depict risk neutral and

Table 11.2. Quality loss estimates in retail per month.

Meat type/pathogen	29% Tolerance	15% Tolerance	10% Tolerance	5% Tolerance	1% Tolerance
Without contamination reduction					
Beef/*Campylobacter*	0.00	0.00	0.00	0.00	451,000.00
Beef/*E. coli*	185,269.98	809,000.00	1,920,000.00	8,130,000.00	8,490,000.00
Chicken/*Campylobacter*	0.00	274,000.00	1,010,000.00	6,020,000.00	7,960,000.00
Chicken/*E. coli*	370,707.69	1,880,000.00	4,670,000.00	20,600,000.00	22,100,000.00
Chicken/*Salmonella*	0.00	0.00	0.00	0.00	138,000.00
Pork/*Campylobacter*	0.00	0.00	0.00	0.00	11,000.00
Pork/*E. coli*	30,181.78	151,000.00	373,000.00	1,630,000.00	1,750,000.00
Turkey/*Campylobacter*	0.00	0.00	0.00	122,000.00	448,000.00
Turkey/*E. coli*	335,937.18	1,870,000.00	477,0000.00	21,500,000.00	23,500,000.00
Turkey/*Salmonella*	0.00	0.00	0.00	69,800.00	347,000.00
With contamination reduction					
Beef/*Campylobacter*	0.00	0.00	0.00	0.00	225,500.00
Beef/*E. coli*	92,634.99	404,500.00	96,0000.00	4,065,000.00	4,245,000.00
Chicken/*Campylobacter*	0.00	137,000.00	505,000.00	3,010,000.00	3,980,000.00
Chicken/*E. coli*	185,353.85	940,000.00	2,335,000.00	10,300,000.00	11,050,000.00
Chicken/*Salmonella*	0.00	0.00	0.00	0.00	69,000.00
Pork/*Campylobacter*	0.00	0.00	0.00	0.00	5500.00
Pork/*E. coli*	15,090.89	75,500.00	186,500.00	815,000.00	875,000.00
Turkey/*Campylobacter*	0.00	0.00	0.00	61,000.00	224,000.00
Turkey/*E. coli*	167,968.59	935,000.00	2,385,000.00	10,750,000.00	11,750,000.00
Turkey/*Salmonella*	0.00	0.00	0.00	34,900.00	173,500.00

strong risk aversion. This wide range of risk attitudes helps to evaluate the robustness of the results.

The results are shown in Table 11.3. The analysis considered the entire set of strategies and tolerance levels for each meat type that could possibly be contaminated with *E. coli* and also for turkey that could possibly be contaminated with *Salmonella*. The other combinations of meat types and pathogens were not relevant because of low or no pathogen prevalence or because there was only one clear strategy for that specific meat and pathogen. The results of the stochastic dominance analysis show that strategies 2 and 3 were cost-effective and highly preferred, except in the case of turkey with possible *E. coli* contamination where the preferred strategy was strategy 1 using a 10% tolerance level. This is a good indication that the current predominant food safety risk reduction strategy (USDA–FDA) implemented by several retail firms across the United States may be less effective than contracting with a private firm or implementing PR/HACCP to mitigate food safety risks at retailing. Understanding incentives or factors to help retail firms implement these more efficient food safety risk reducing strategies are then evaluated using the MICMAC scenario method.

Table 11.3. Stochastic dominance comparison of intervention strategies.

Meat type/pathogen	Preferred strategy and tolerance level	
	RAC=0.000001	RAC=0.1
Beef/*E. coli*	Strategy 3 @ 1%	Strategy 3 @ 1%
Chicken/*E. coli*	Strategy 3 @ 5%	Strategy 3 @ 5%
Pork/*E. coli*	Strategy 2 @ 1%	Strategy 2 @ 1%
Turkey/*E. coli*	Strategy 1 @ 10%	Strategy 1 @ 10%
Turkey/*Salmonella*	Strategy 3 @ 5%	Strategy 3 @ 5%

Summary

Scenario analysis is a statistical procedure used to determine what factors or variables would be instrumental in implementing a desired outcome. Instead of forecasting what a particular outcome will be, scenario analysis allows you to envision the desired outcome and then determine what factors will lead to that outcome.

Stochastic dominance analysis suggests a need to extend the PR/HACCP performance standard to more retail facilities and other pathogens. Currently, *Salmonella* is the only performance standard for

PR/HACCP. The assumption is that if *Salmonella* levels decrease, then other levels of pathogens will follow. This might not be true with retailing.

References

Antle, J.M. (2000) No such thing as a free safe lunch: the cost of food safety regulation in the meat industry. *American Journal of Agricultural Economics* 82, 310–322.

Burton, S., Howlett, E., and Tangari, A.H. (2009) Food for thought: how will the nutrition labeling of quick service restaurant menu items influence consumers' product evaluations, purchase intentions, and choices? *Journal of Retailing* 85, 258–273.

CDC (2006) Multistate outbreak of *E. coli* O157 infections linked to Taco Bell (FINAL UPDATE). Available at: http://www.cdc.gov/ecoli/2006/taco-bell-12-2006.html (accessed 17 October 2016).

CDC (2011) Estimates of foodborne illness in the United States. United States Centers for Disease Control. Available at: http://www.cdc.gov/foodborneburden/2011-foodborne-estimates.html (accessed 30 December 2015).

CIDRAP (2008) Foodborne disease rates changed little in 2007. Center for Infectious Disease Research and Policy. Available at: http://www.cidrap.umn.edu/news-perspective/2008/04/foodborne-disease-rates-changed-little-2007 (accessed 30 December 2015).

Grewal, D., Baker, J., Levy, M., and Voss, G. (2003) The effects of wait expectations and store atmosphere evaluations on patronage intentions in service-intensive retail stores. *Journal of Retailing* 79, 259–268.

Godet, M. (1993) *From Anticipation to Action*. UNESCO Publishing, Paris.

Hosking, G. (1984) Quality cost measurement in the food industry. *Food Technology (in Australia)* 36, 165–167.

Kay, S. (2003) $2.7 Billion: the cost of *E coli* O157:H7. *Meat & Poultry* February, 26–34.

Kerin, R., Jain, A., and Howard, D.J. (1992) Store shopping experience and consumer price-quality-value perceptions. *Journal of Retailing* 68, 376–397.

Knutson, R.D., Cross, H.R., Acuff, G.R., Russell, L.H., Boadu, F.O., Nichols, J.P., Wang, S., Ringer, L.J., Childers, A.B., and Savell, J.W. (1995) *Reforming Meat and Poultry Inspection: Impacts of Policy Options*. Institute for Food Science and Engineering, Agricultural and Food Policy Center, Center for Food Safety, A&M University, TX, USA.

Lupin, H.M., Parin, M.A., and Zugarramurdi, A. (2010) HACCP economics in fish processing plants. *Food Control* 21, 1143–1149.

Marler, C. (2010) HACCP: managing food safety, food poison journal, food poisoning outbreaks and litigation surveillance and analysis. Available at: http://www.foodpoisonjournal.com/ (accessed 17 October 2016).

McKinley, J. (2006) Center of *E. coli* outbreak, center of anxiety. Available at: http://www.nytimes.com/2006/09/25/us/25ecoli.html (accessed 30 December 2015).

Pratt, J.W. (1964) Risk-aversion in the small and in the large. *Econometrica* 32, 122–136.

Taguchi G. (1986) *Introduction to Quality Engineering. Designing Quality into Products and Processes*. Asian Productivity Organization, Tokyo.

Weise, E. and Schmit, J. (2007) Spinach recall: 5 faces, 5 agonizing deaths. 1 year later. Available at: http://a.abcnews.com/Business/Story?id=3633374&page=1 (accessed 15 December 2016).

Zugarramurdi, A., Parin, M.A., and Lupin, H.M. (1995) Economic engineering applied to the fishery industry. *FAO Fisheries Technical Paper*, 351. Food and Agriculture Organization, Rome.

12 Cost of Microbial Foodborne Outbreaks to Society

Key Questions
- What are the categories of food safety costs to society?
- How do we quantify the cost to society of foodborne illness and death?
- How do we estimate the net loss to all sectors?
- How do food safety risks and perception of food risks affect the cost to society?
- How can food policy have unintended consequences and increase the cost to society?

The Scope of Food Safety Cost to Society

The cost of food safety risks goes beyond the firm level of costs discussed in Chapter 11. Food safety costs to society also include:

1. Detailed cost of illness or death to consumers, which could be included as part of firm failure costs. This cost category includes medical costs, productivity loss from not being able to work for a period of time, and estimates of premature death.
2. Food safety risks could also result in food scares and negative consumer perceptions of our food supplies. Negative consumer perceptions could, in turn, lead to decreased food demand and loss of businesses and employment, which could severely impact the food industry and regional or national economies. Loss of consumer demand could also extend trade embargoes and losses (see Chapter 14).
3. The final category of cost includes poorly designed policies. Inefficient policy design could result in free rider problems and offsetting behavior (unintended consequences) that increase the cost to society.

The cost of microbial foodborne outbreaks to society is a growing concern around the globe. These costs are in the billions of dollars in the United States alone. Losses could also cause complete market failure and outrage when minimum safety standards are not enforced by the private and public sectors.

Cost to Society of Foodborne Illness and Death

The United States Department of Agriculture's Economic Research Service (USDA-ERS) provides estimates of the benefits or reductions in illness caused by four foodborne pathogens: *Salmonella*, *Escherichia coli* O157:H7, *Campylobacter jejuni* or *C. coli*, and *Listeria monocytogenes*. USDA-ERS provides a range of estimated benefits from potential low to high values. Four reasons suggested why a range should be used:

1. The incidence of foodborne illness (and death) and the proportion of cases caused by contaminated meat and poultry were uncertain.
2. The efficacy of the hazard analysis and critical control points (HACCP) program in reducing foodborne illness was also uncertain. The highest benefit estimate reported incorporates an efficacy rate of 90%, while the lowest estimate uses a rate of 20%.
3. The benefit estimates varied because two different discount rates were used. Lower estimates used a relatively high discount rate of 7% to reflect private valuations, while the higher estimates used a discount rate of 3% to reflect a societal viewpoint.
4. There was use of two different methods to assign economic value to improvements in health and longevity resulting from reductions in foodborne illness. The higher benefit estimates reported used

the willingness to pay, while the lower used a variant of the cost-of-illness methodology.

Tables 12.1, 12.2, 12.3, and 12.4 show detailed cost categories from USDA-ERS cost simulators for four main pathogens: *E. coli, Campylobacter, Listeria*, and *Salmonella*. The highest cost is obviously premature death, followed by medical costs, and then productivity loss. In 2013, the estimated total loss for *Salmonella* was US$3.66 billion, the highest loss of all foodborne pathogens. *E. coli, Campylobacter* and *Listeria* had comparable loss estimates.

A major challenge we face today is that, even though the number of food recalls could be declining due to more stringent control measures, the magnitude of and average loss continue to be significantly larger. One possible reason is the increased volumes and velocity of food flows and trade.

Estimating Net Loss to All Sectors

The costs of foodborne illness to society are more complex than the value of illness or death or costs to the firm. As ironic as it may sound, some sectors will benefit from a food recall incident while other sectors and individuals will suffer loss. For example, the medical industry could benefit from increased patients' bills from having more sick people. To estimate the cost of foodborne illness, we therefore need to estimate costs to individuals and firms and benefits to other industries and sectors. One approach to estimate the cost to society is to use the social accounting matrix (SAM). SAM could also be used to estimate cost-effectiveness of cost mitigation strategies like HACCP. Studies with SAM show that for every US$1 of income saved by preventing foodborne illness, the overall economy saves US$1.92 (Golan *et al.*, 2000). So, how does the SAM approach operate?

SAM helps us quantify costs and benefits of foodborne illness and death to different sectors of the economy. SAM is a snapshot view of the circular flow of accounts in an economy (Golan *et al.*, 2000). It is an input–output (I–O) model with expenditures by some sectors and receipts to other sectors. Golan *et al.* (2000) noted that SAM comprises of a set of production activities (such as meat and poultry processors), commodity markets for goods and services (such as meat and poultry products), factors (labor and capital), households, a capital account, and other institutions (government

and the rest of the world). A detailed explanation of SAM composition can be accessed at http://ageconsearch.umn.edu/bitstream/34023/1/ae000791.pdf. Numbers presented in the subsections below relate to SAM estimates.

Which household type receives the most net benefit?

The SAM model analyzed two different kinds of benefits. The final economy-wide distribution of the benefits of fewer illnesses and premature deaths differed from the initial distribution of benefits. Initially, the benefits of these reductions accrued to those who would have fallen sick or would have died prematurely.

However, unlike the initial distribution of benefits, the final distribution did not mirror disease incidence, but depended instead on the relationship of households to the economy. As a result, higher income households, which have strong links to the economy, bore a larger share of the change in economic activity triggered by reduced premature deaths and medical expenses than lower income households, which have weak links to the economy.

How do HACCP costs compare to reduction in premature deaths?

The reduction in premature deaths initially resulted in increased household income. The study used the human capital approach, and the reduction in premature deaths resulting from HACCP translated into an increase in national income. The increase was calculated as US$5.25 billion, which was distributed among different households (e.g. with children, without children, above poverty, below poverty, and elderly).

This initial increase in national income did not represent the ultimate impact, because households responded to the initial increase in income by expanding consumption and savings. This expansion triggered further increases in economic activity extending far beyond the originally affected households.

The difference between initial and final benefits of premature death is quite interesting. Households with children gained a smaller percentage of benefits in the final benefits distribution than in the initial distribution, whereas childless households and elderly-headed households gained a higher percentage. In fact, although elderly-headed households were not allocated any initial benefits of reductions in premature

Table 12.1. Cost of foodborne illness estimates for *Escherichia coli* O157.

E. coli O157	Total cases	Non-hospitalized		Hospitalized				
		Didn't visit physician; recovered	Visited physician; recovered	Hospitalized, non-HUS; recovered	Hospitalized, HUS only; recovered	Hospitalized HUS & ESRD;[a] survived with later premature death	Hospitalized, non-HUS; died	Hospitalized, HUS; died
Number of cases								
Low	17,587	13,769	3,269	468	78	3	0	0
Mean	63,153	49,278	11,737	1,806	302	10	8	12
High	149,631	117,208	27,809	3,838	642	21	45	68
Cost component								
Medical costs (average cost per case)		US$3	US$420	US$9,128	US$55,617	US$947,411	US$9,119	US$55,613
Medications		US$3	US$18	US$53	US$31	US$43	US$44	US$27
Office visits			US$173	US$139	US$262	US$262	US$139	US$262
Emergency room visits			US$229	US$504	US$728	US$728	US$504	US$728
Hospitalization				US$8,433	US$54,596	US$54,596	US$8,433	US$54,596
Chronic medical						US$891,781		
Productivity loss, non-fatal cases		US$29	US$211	US$614	US$75	US$58,890	US$960	US$369
Premature death								
Low value per death						US$1,027,644	US$1,574,065	US$1,574,065
Mean value per death						US$5,800,852	US$8,657,357	US$8,657,357
High value per death						US$10,276,443	US$15,740,649	US$15,740,649

[a]The assumption is that end-stage renal disease (ESRD) shortens life expectancy. Thus, these cases are counted as resulting in deaths at a future date, the value of which is discounted back to present value.

HUS, hemolytic–uremic syndrome.

Based on Hoffman *et al.* (2012) and Batz *et al.* (2014).

Table 12.2. Cost of foodborne illness estimates for *Campylobacter* (all species).

Health outcomes	Total acute cases	Acute illnesses					Chronic illnesses	
		Non-hospitalized		Hospitalized	Post-hospitalization outcomes			
		Didn't visit physician; recovered	Visited physician; recovered	Hospitalized	Post-hospitalization recovery	Hospitalized; died	Morbidity from Guillain–Barré syndrome	Deaths associate with Guillain–Barré syndrome
Number of cases								
Low	337,031	314,531	18,200	4,300	4,300	0	1,485	65
Mean	845,024	790,930	45,631	8,463	8,387	76	1,916	86
High	1,611,083	1,508,858	86,998	15,227	14,895	332	2,602	111
Cost component								
Medical costs								
Physician office visits								
Average visits per case		0	1.4	0.7	1	1		
Average cost per visit		US$136	US$136	US$136	US$136	US$0		
Emergency room visits								
Average visits per case		0	0.1	0.3	0			
Average cost per visit		US$573	US$573	US$573	US$573	US$0		
Outpatient clinic visits								
Average visits per case		0	0.3	0.2	0			
Average cost per visit		US$659	US$659	US$659	US$659	US$0		
Hospitalizations								
Average admissions per case		0	0	1	0			
Average cost per hospitalization		US$13,938	US$13,938	US$13,938	US$13,938	US$0		
Total medical costs per case		US$0	US$445	US$14,337	US$136			
Productivity loss, non-fatal cases								
Proportion of cases per employee		0.444596	0.458895	0.430292	0.430292	0.430292		

Continued

Table 12.2. Continued.

Health outcomes		Per case assumptions, 2014 (in 2013 dollars)						
		Acute illnesses					Chronic illnesses	
		Non-hospitalized		Hospitalized	Post-hospitalization outcomes		Morbidity from Guillain–Barré syndrome	Deaths associate with Guillain–Barré syndrome
	Total acute cases	Didn't visit physician; recovered	Visited physician; recovered	Hospitalized	Post-hospitalization recovery	Hospitalized; died		
Average number of work days lost		0.5	1.5	4.5	2.25			
Average daily earnings		US$254	US$256	US$263	US$263			
Productivity costs per case		US$57	US$176	US$509	US$254			
Premature death								
Low value per death						US$1,574,065		US$1,574,065
Mean value per death						US$8,657,357		US$8,657,357
High value per death						US$15,740,649		US$15,740,649
Chronic: Guillain–Barré syndrome								
Medical costs and productivity loss per case							US$167,232	

Based on Hoffman *et al.* (2012) and Batz *et al.* (2014).

Table 12.3. Cost of foodborne illness estimates for *Listeria monocytogenes*.

Per case assumptions, 2013 (in 2013 dollars)

Cost Component	Total cases	Non-congenital total cases	Non-hosp.: Didn't visit physician; recovered	Non-hosp.: Visited physician; recovered	Hospitalized, maternal	Hospitalized, other adults, moderate	Hospitalized, other adults, severe (ICU), recovered	Hospitalized, other adults, severe, died	Hospitalized, newborn, total	Hospitalized, newborn, recovered	Newborn deaths	Stillbirths	Newborn, mild disability	Newborn, moderate to severe disability	Newborn, total disability
Number of cases															
Low		456	36	420	70	12	338	0	101	71	0	18	2	7	2
Mean	1,591	1,309	136	1,173	196	33	697	247	282	189	8	51	7	20	7
High	3,161	2,577	143	2,434	407	68	1,248	710	584	385	23	105	14	42	14
Cost Component															
Medical costs															
Physician office visits															
Emergency room visits															
Outpatient clinic visits															
Hospitalizations															
Average cost per case, regular hospitalization					US$32,760	US$32,760	US$32,760		US$65,520	US$65,520	US$65,520	US$65,520	US$65,520	US$65,520	US$65,520
Average cost per case, ICU							US$65,520	US$65,520	US$65,520	US$65,520	US$65,520	US$65,520	US$65,520	US$65,520	US$65,520
Average cost per case of hospitalization					US$32,760	US$32,760	US$98,280	US$65,520	US$131,040	US$131,040	US$131,040	US$131,040	US$131,040	US$131,040	US$131,040
Productivity loss, non-fatal cases															
Average number of work days lost					15.0	15.0	30.0								
Productivity loss per case					US$1,879	US$1,154	US$2,308								
Premature death															
Low VSL per death								US$1,574,065			US$1,574,065	US$1,574,065			
Mean VSL per death								US$8,657,357			US$8,657,357	US$8,657,357			
High VSL per death								US$15,740,649			US$15,740,649	US$15,740,649			
Chronic Illnesses															
Medical and special education costs per case													US$105,669	US$264,174	US$1,447,601
Productivity loss per case													US$458,903	US$1,580,664	US$1,699,638

ICU, intensive care unit; VSL, value of statistical life.
Based on Hoffman et al. (2012) and Batz et al. (2014).

Table 12.4. Cost of foodborne illness estimates for *Salmonella* (non-typhoidal).

| Health outcomes | Total cases | Non-hospitalized | | Mean estimates, 2013 | | Post-hospitalization outcomes | |
| | | Didn't visit physician; recovered | Visited physician; recovered | Hospitalized | | Post-hospitalization | Hospitalized, died |
				Hospitalized	Post-hospitalization		
Number of cases	1,027,561						
Cases by outcome		934,241	73,984	19,336		18,958	378
Medical costs							
Physician office visits			US$14,082,149	US$1,840,202		US$2,577,468	
Emergency room visits			US$4,238,820	US$3,323,478		US$0	
Outpatient clinic visits			US$14,622,881	US$2,547,817		US$0	
Hospitalization			US$0	US$269,505,637		US$0	
Total medical costs by outcome			US$32,943,851	US$277,217,134		US$2,577,468	
Premature death							US$32,272,480,959
Productivity loss, non-fatal cases		US$52,810,181	US$13,911,195	US$9,836,930		US$4,822,314	
Total cost by outcome		US$52,810,181	US$46,855,046	US$287,054,064		US$7,399,782	US$32,272,480,959
Total cost of illness	US$3,666,600,031						

Based on Hoffman *et al.* (2012) and Batz *et al.* (2014).

deaths, they received 6% of the final benefits. These differences arose because, unlike the initial distribution of benefits, the final distribution did not mirror disease incidence, but depended instead on the linkages between households and the economy. A similar pattern appears when households above and below poverty are compared. Poor households realized 17% of the initial increase in income due to reductions in premature deaths, but only 9% of the final increase, reflecting the fact that lower income households have weaker factor-payment linkages to industrial production than other households. Conversely, upper income households with strong factor-payment linkages were more strongly affected by changes in the returns to labor and capital.

How much will the medical sector lose with HACCP implementation and why?

While tracing the benefits, the SAM model also analyzed the economic impact of reductions in direct medical expenses when these expenses are paid by health insurance. The National Health Interview Survey identified three different households: (i) households with public coverage; (ii) private coverage; and (iii) households with no coverage. This classification distinguished households whose health care costs were wholly or partially subsidized by public programs from households protected by private insurers and households lacking any kind of coverage.

Public coverage took precedence in the classification in order to identify all households receiving public funds. The majority of non-elderly households fell into the private insurance category; 65% of households with children and 74% of households without children had private coverage. In contrast, elderly households depended almost exclusively on public health insurance coverage, reflecting the role of Medicare in providing health care for the elderly.

The distribution of illness by household insurance category was US$4.92 billion dollars in medical expenditure savings. Households with private coverage accrued a much larger share of total savings (US$3.19 billion) than households with public coverage (US$1.39 billion) and households without coverage (US$0.34 billion). Thus, the availability of health insurance changes the linkages examined. Most important was that nearly one-third of medical expenses were incurred by households with coverage or no coverage.

The SAM model traced the impact of reductions in direct medical costs when third-party payers (private insurers or the government) paid the bills. The initial drop in medical expenses for publicly insured and uninsured households was deducted from medical sectors and distributed back to households as "tax cuts." Specifically, the US$1.73 billion reduction in the medical expenses of publicly insured and uninsured households was distributed back to households above poverty. These households increased their consumption and savings accordingly.

The initial impact of the reduction in medical costs for privately insured households was represented by a US$3.19 billion decrease in costs for the insurance sector, by diverting insurance sector expenditures from the purchase of medical goods and services to the purchase of other goods and services, as indicated by the expenditure coefficients in the SAM.

Perception of Food Risks and Cost to Society

In this section, we discuss the concept of perceptions of food risks and illustrate how consumers react to food scares. It is important to note that risk perception and food scares could have greater cost impacts than implicated food recalls. These could impact long-term demand. Food scares have been defined as "single or collective incidents, which are particularly focused upon by the media and by relevant Government agencies" (Knowles *et al.*, 2007). Food scares are often characterized by a sudden escalation of consumer anxiety as the public assesses the magnitude of risk they face from a food safety incident and make consumption decisions in the presence of risk and uncertainty. Upheaval in purchasing patterns for a certain food can result in a shift of market demand for the food (Jin *et al.*, 2003).

Food is an essential need. Individuals must make decisions about food selection every day and, as in most important decisions, the risk related to food choices must be considered. Food safety events compound the risk associated with food-selection decisions. Previous research has looked at how the public assesses food safety information it receives about food. Studies have shown that food-purchasing behavior is affected as shoppers react to uncertainty in food safety and seek to reduce the perceived health risks of consuming potentially contaminated food products by reducing or eliminating exposure

(Huang, 1993; Verbeke, 2001; Smed and Jensen, 2005). Predicting the level of consumer response is complicated by the public's tendency to overestimate some risks while underestimating others, distorting the relationship between the actual level of risk in a food scare and the perceived risk to the consumer (Sandman, 1987; Verbeke *et al.*, 2007).

Occurrence of food scares in the United States and the United Kingdom

Food safety is a topic in the front of the mind for many American and British consumers in light of the increasing number of reported food safety incidents. In the first 9 months of 2007, there were 41 food recalls overseen by the USDA Food Safety and Inspection Service (FSIS) in comparison to 28 such recalls during the same time period in 2006 (FSIS, 2007).

In the United Kingdom, the Food Standards Agency (FSA) has been charged with protecting public health and consumer interests in relation to food. Although FSA reported a 19.2% reduction in foodborne illnesses between 2000 and 2006, there were still more than 53,000 laboratory-reported cases of foodborne pathogens in the United Kingdom in 2005 (FSA, 2007). In 2006, the Agency conducted 1,342 investigations of food-related incidents, including high-profile cases like a *Salmonella* outbreak associated with chocolate, benzene contamination of soft drinks, and the unauthorized genetically modified organism contamination of rice in the United States.

The United States Centers for Disease Control and Prevention (CDC) and its predecessors have been tracking foodborne and waterborne diseases for more than 60 years. Investigating foodborne disease outbreaks has resulted in the enactment of important public health measures, such as the Pasteurized Milk Ordinance, which has decreased the incidence of illness and death related to pasteurized foods. In 2000, a CDC report looked at foodborne disease outbreaks reported between 1993 and 1997. During this period, 2,751 outbreaks were reported, which resulted in 86,058 persons becoming ill and 29 deaths. For the majority of outbreaks where a cause could be determined, 75% of the outbreaks and 86% of the cases were caused by bacterial pathogens. The most common pathogen was *Salmonella* serotype Enteritidis, consumed primarily in raw or undercooked eggs. It is important to note that the number of outbreaks summarized above is a small percentage of the actual outbreaks during that timeframe. The majority of outbreaks go unrecognized, especially because the lag between ingestion and illness can be 1 to 5 days. Those outbreaks that are recognized often go unreported to public health authorities because of the complicated chain of events that must occur for an outbreak to be reported to the CDC (Olsen *et al.*, 2000).

Meanwhile, food scares can have a far-reaching impact on producers that export their food products overseas. One such situation was the December 2003 identification of the first case of "mad cow" disease in the United States. Virtually all United States beef exports were halted as major trading markets were promptly closed to United States beef. It was estimated that by late 2007, United States beef producers, feed lots, and processors had lost nearly US$12.5 billion in revenue alone in just the Japanese and South Korean markets since the bans were enacted (Doering, 2007). In addition to lost revenue, cattle producers are also faced with implementing export-verification programs to meet the newly created trade rules imposed by the importing countries in reaction to the perceived food dangers. It has been anecdotally estimated to cost US$8,000 for a cattle producer to develop a USDA-approved and -audited Quality Systems Assessment program. This estimate does not include the cost of mandatory ongoing audits (Mark and Halstead, 2007). It should be noted that food safety occurrences and losses are similar in the United States and United Kingdom, and strategies to mitigate these risks could be easily shared.

Risk perception and consumer reaction to food scares

The measurement and effect of risk perception in consumption decisions have been a frequent research topic. Pennings *et al.* (2002) suggested that looking at a consumer's risk perception and risk attitude can help in the prediction of consumer reaction to food safety risks. Risk perception represents an individual's perceived likelihood of exposure to the food safety risk. Risk attitude represents an individual's general sensitivity to risk. It is anticipated that the greater the risk consumers associate with a product, the more significant the behavioral response to mitigate risk.

Wansink (2004) identified four segments of consumers defined by their overall pre-existing attitudes toward risk and their perception of the specific food safety risk.

1. The *accountable* segment are those who have a low aversion to risk and a low perception of the specific food safety risk presented. They see themselves as responsible for their behaviors and any resulting consequences.

2. The *concerned* segment also has a low aversion to risk but they see the chance of the specific food safety risk as high. Since they had a low pre-existing attitude toward risk in general, it is assumed that their behavior regarding the food safety matter is resulting from their perception of the risk at hand.

3. The *conservative* segment is highly risk averse and will avoid unnecessary risk.

4. The *alarmist* segment tends to overreact to risk, as they are averse to risk and perceive personal danger in the specific food safety risk.

Using this framework, it is expected that those consumers who perceive no risk from the food safety situation will not change their behavior regarding the food product. However, if there is a perceived personal risk, the behavior toward consumption of the product in question will additionally be affected by the consumer's general attitude toward risk. Risk-averse consumers who also perceive a personal risk would demonstrate the largest decrease in the behaviors that might expose them to the product.

Schroeder *et al.* (2007) took the concept of risk perceptions and attitudes and created a consumer survey to measure the magnitude of these factors. Their resulting survey of 4,000 consumers found risk perceptions had more influence than risk attitudes when assessing food safety concerns with beef. In this case, a person's negative perception of the food safety resulted in a reduced willingness to consume beef, and a negative risk attitude produced a stronger disagreement that eating beef was worth the risk.

In the Schroeder *et al.* (2007) study, the magnitude of the consumer reaction varied across countries represented in the survey. This phenomenon was also demonstrated in the study by Pennings *et al.* (2002), where cultural differences in risk attitudes and risk perceptions were noted. In this study, German, American, and Dutch consumers were surveyed about beef consumption behaviors related to bovine spongiform encephalopathy (BSE). It was found that the German consumers, who as a group demonstrated the highest level of risk attitude and risk perception, reacted most strongly to the BSE crisis. This difference in the sensitivity to risk from BSE resulted in different behaviors among consumers of different countries who were actually at similar risk for contracting Creutzfeldt–Jacob disease from eating contaminated beef. German consumers expressed more concerns about eating beef, with nearly 60% of those surveyed reporting a decrease in beef consumption because of the BSE crisis. By comparison, in the Netherlands and in the United States – countries where the average consumer reported lower responsiveness to BSE risks than in Germany – 23% and 18%, respectively, of those surveyed reported a decrease in beef consumption related to the BSE scare.

Media influence, demographic factors, and consumer reaction to food scares

When assessing the risk of a food scare, most consumers turn to the media for insight into the unfolding events. Whether through television, radio, or print, it is clear that many consumers depend on the media for information about food safety and how it impacts them. A model developed by Richards and Patterson (1999) showed that while positive and negative media attention have the anticipated associated change in price for the product, it is clear that the magnitude of the consumer reaction is greater with negative publicity. The finding that consumers place more weight on negative media reports was also reported in research regarding the 1982 milk contamination scare in Hawaii. The indication that positive media messages carry less weight with consumers is significant to governments and food industry firms who are interested in reassuring the public about the safety of the food supply (Smith *et al.*, 1988).

Verbeke and Viaene (1999) analyzed television coverage of meat issues in Belgium in the mid-1990s. They estimated that more than 90% of the broadcast meat messages portrayed a negative relationship between fresh meat safety and human health. This research indicated that the more attention a consumer reported to have paid to the media coverage, the more likely the consumer was to reduce meat consumption. In research done on United States media articles and their impact on meat demand, Piggott and Marsh (2004) determined that the reduction in demand for an affected meat product following publication of food safety information was limited to a contemporaneous effect. While major food safety events, measured by the number of newspaper articles related to the event, caused significantly larger demand responses, there appeared to be only a minor long-run impact on demand.

Demographics may also play a part in consumer reaction. Smed and Jensen (2005) took a look at yet another food scare: *Salmonella* in shell eggs. This research also focused on how negative media attention about a food scare can affect consumer purchasing behavior. In this study, researchers considered the amount of negative radio and newspaper coverage regarding shell egg contamination in a given week and then reviewed consumer-reported purchase data for a "safe" alternative product – pasteurized eggs – during the same time period. Results from this study showed significant differences in the magnitude of the shift toward purchasing pasteurized eggs during the scare for various groups of consumers. Age and education were two important factors, with the aged and those with low education levels showing the greatest response. Similar demographic influences related to age were reported in the Verbeke *et al.* (2000) study, where researchers found that households with older adults or children were more likely to decrease their meat consumption after the BSE crisis; however, they found no such effect associated with level of education of the consumer.

Pricing strategies and consumer reaction to food scares

A relatively new area of research is the mitigating effect retail strategies can have on consumer purchase behavior when shoppers are faced with the decision to purchase food items that have been associated with food safety concerns. A recent food safety topic for consumers has been the apprehension over genetically modified (GM) food. The negative reaction in Europe has been so strong that from 1998 until 2004 the European Union (EU) was not approving new GM foods. The EU removed its moratorium on GM foods in 2004 and instituted new labeling and traceability requirements for all GM food products beyond a 0.9% tolerance level. EU retailers have been unwilling to add GM products to their shelves because of the negative consumer sentiment, although no alleged health or environmental hazards have been scientifically proven (Moon *et al.*, 2004). Even with this assumed overarching rejection of GM food by the public, research has shown that in the United Kingdom nearly 47% of consumers surveyed would be willing to buy GM food products. It is important to note that most of these respondents required a discount from the price of the equivalent non-GM product as an incentive to make such a purchase (Moon *et al.*, 2004).

Previous research has considered both risk profiles and purchase behavior when assessing consumer reaction to food scares. While self-reporting of risk perceptions and attitudes is the only way to know how consumers are feeling about risk, survey data that require consumers to recall or predict changes in behavior can be less reliable. We provide an empirical construct of a consumers' risk perception index.

The majority of prior research on consumer reaction has been limited to one of two types. Qualitative research dependent on surveys to collect data regarding consumer behaviors and attitudes relating to food safety has been common (Verbeke and Viaene, 1999; Pennings *et al.*, 2002; Schroeder *et al.*, 2007). This research tends to be limited, as surveys are of narrow groups of subjects in particular regions of a country or selective markets. Additionally, surveys are dependent on self-reporting of past and anticipated behaviors, an approach that can sometimes result in respondent bias (Lee *et al.*, 2000).

The other approach is that of using event models in which statistical analysis is performed on data gathered after a significant food safety event occurs. In these cases, the research is looking at changes in the broad market and associating those with the timing of the incident. Examples of this type of research include Thomsen and McKenzie (2001) and Piggott and Marsh (2004).

In either case, past research has considered consumers as a homogeneous group. For example, when sales of a product went down at the time of a food safety incident, it was assumed that all consumers reduced their purchases of the product in reaction. Researchers have used actual supermarket purchase data that are linked with customer loyalty card demographic information. This combination has allowed researchers to track consumer purchase habits before, during, and after a food safety incident and record the heterogeneity of the behaviors based on consumer segmentation across all regions of the country using genuine purchase data that are not affected by respondent bias. We also analyze price promotion strategies that can mitigate risk related to outrage or dread rather than the actual hazard.

Risk perception construct

It has been suggested by Sandman (1987) that perceived risk is equal to actual hazard plus outrage (the unknown), so that levels of perceived risk and levels of actual risk are often uncorrelated as a result of outrage. Risk as perceived by the public is often more a condition of outrage than the actual hazard. While the hazard can be low, economic loss can be significant if consumers assign high levels of outrage when assessing the risk. In determining an individual's response to risk, there are outrage factors that affect how one judges the magnitude of the risk. Varying degrees of the outrage experienced by different consumers facing the same food safety incident could result in a heterogeneous perception of risk.

Table 12.5 shows the categorization of the risk perceptions and risk attitudes index (Nganje and Wolf, 2008). Each section of the quadrant is defined by overall pre-existing attitudes toward risk and their perception of the specific food safety risk.

In Table 12.5, the *accountable* section is for those who have a low aversion to risk and a low perception of the specific food safety risk presented. They see themselves as responsible for their behaviors and any resulting issues. The life-stage segment Young Adult has been matched with this section because members of this segment commonly demonstrate a sense of invincibility and disengagement with food, in terms of its credence attributes, which would be reflected in a low risk attitude and risk perception.

The *conservative* section of the quadrant is for those who are highly risk averse and will avoid unnecessary risk, yet are not acutely sensitive to food safety concerns. Older families exhibit behavior that places them in the *conservative* section where, although they are sensitive to risks facing their increasingly independent children and may

have anxiety over increased personal obligations, they are less concerned about passing food safety issues that are unlikely to affect them.

Those individuals associated with the *concerned* segment also have a low aversion to risk but they see the chance of the specific food safety risk as high. Since they have a low pre-existing attitude toward risk in general, it is assumed that their behavior regarding the food safety matter is resulting from their perception of the risk at hand. Older adults fall in the *concerned* quadrant as their trepidation over day-to-day concerns is moderated by life experience and absence of care for children in the household, while health concerns that could be intensified by foodborne illnesses make them more sensitive to food safety risks.

The final section of the quadrant is categorized as *alarmist* and identifies those who tend to overreact to risk, as they are averse to risk and perceive personal danger in the specific food safety risk. Young families and pensioners show the highest levels of risk attitude and perception as they are facing periods of their lives when health issues are a high priority and there are general concerns about maintaining personal safety either for children or themselves.

Using this framework, it is expected that those consumers who perceive no risk from the food safety scare will not change their behavior regarding food purchase (Nganje and Wolf, 2008). Segments meeting this designation are Accountable/Young Adults and Conservative/Older Families. For those who have a high sensitivity to food safety risk, the degree of reaction is influenced by the attitude about general risk. It is expected that risk-averse consumers who also perceive a personal risk, the Alarmist/Young Families and Alarmist/Pensioners, would demonstrate the largest decrease

Table 12.5. Life-stage segments defined within risk quadrant.

		Risk attitude	
		Low	High
Risk perception (of food safety risk)	Low	*Accountable* Young adults/brand familiarity	*Conservative* Older families/brand familiarity
	High	*Concerned* Older adults/brand familiarity	*Alarmist* Young families/brand familiarity Pensioners/brand familiarity

in the behaviors that might expose them to the affected product, while the reaction by the Concerned/Older Adults would be less.

Length of food scare and economic loss

The length of the scare considered in this research was determined by the concentration and duration of the British newspaper coverage immediately following the avian flu outbreak announcement of 3 February 2007 (we have used avian flu as an example because of data availability; however, similar patterns could be observed in foodborne pathogens). Headlines regarding the outbreak were recorded for *The Daily Telegraph*, *The Guardian*, *The Independent*, and *The Sun* newspapers. Figure 12.1 shows the total number of articles from these publications that were published on each day of the scare. As is shown in Figure 12.1, the highest level of newspaper coverage occurred in the 3 weeks following the announcement. These 3 weeks are contained in the 4-weekly sales periods starting 29 January, 5 February, 12 February, and 19 February, 2007. Within 3 weeks of the announcement, coverage of the outbreak had dwindled from daily reporting of public health concerns to occasional stories regarding the economic effects of the food scare, including the loss of jobs and reduction in sales for Bernard Matthews Ltd.

Perception of food risks: The case of avian flu

On 1 February 2007, a veterinarian was called to the Bernard Matthews farm in Holton, Suffolk, in eastern England to begin an investigation into the deaths of 2,600 turkeys. Bernard Matthews Ltd is the top turkey processor in the United Kingdom operating 57 turkey farms in Norfolk, Lincolnshire, and Suffolk. Started in 1950 by company chairman and spokesman, Bernard Matthews, the firm was publicly traded from 1970 to 2000. After a takeover attempt by the American company Sara Lee, the firm returned to private ownership by the family. It is estimated that Bernard Matthews Ltd has more than 7,000 employees at farms producing more than 8 million turkeys annually.

On 3 February 2007, the European Commission said tests confirmed that the birds on the Bernard Matthews Holton farm were infected with the lethal H5N1 strain of avian influenza (BBC News, 2008). While the United Kingdom has experienced only three outbreaks of H5N1 avian flu, thousands of outbreaks have been recorded internationally. Table 11.6 shows an alphabetical list of the 48 countries that have reported H5N1 avian flu outbreaks affecting domestic poultry from 2003 to 2008 with the number of outbreaks recorded during this time.

The outbreak of February 2007 became the largest European avian flu event, with nearly 160,000 birds culled. This was the first time that avian flu had been linked to a prominent consumer brand name (Smith, 2007). Although the government and Bernard Matthews Ltd released statements assuring the public that poultry products were safe when cooked properly, a drop in sales was reported as consumers reacted to the news. Figure 12.2 shows the impact in both sales volume (units sold) and

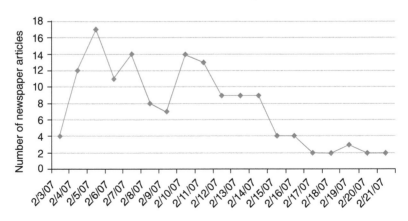

Fig. 12.1. Number of newspaper articles regarding the avian flu scare by date of publication. Adapted from Cacciolotti *et al.* (2007).

Table 12.6. Countries[a] reporting H5N1 avian flu in domestic poultry, 2003 to 2008.

Country	Number of incidents	Country	Number of incidents
Afghanistan	22	Kazakhstan	1
Albania	3	Korea	59
Azerbaijan	2	(Republic of)	
Bangladesh	285	Kuwait	20
Benin	5	Laos	10
Burkina	4	Malaysia	16
Faso		Myanmar	93
Cambodia	20	Niger	2
Cameroon	1	Nigeria	61
China	93	Pakistan	50
Côte d'Ivoire	4	Palestinian	8
Czech	4	Autonomous	
Republic		Territories	
Denmark	1	Poland	10
Djibouti	1	Romania	163
Egypt	1,065	Russia	147
France	1	Saudi Arabia	29
Germany	7	Serbia and	1
Ghana	6	Montenegro	
Hungary	9	Sudan	18
India	48	Sweden	1
Indonesia	261	Thailand	1,139
Iran	1	Togo	3
Iraq	3	Turkey	219
Israel	10	Ukraine	42
Japan	9	United Kingdom	3
Jordan	1	Vietnam	2,475

[a]Hong Kong reported H5N1 in domestic and wild birds in January, 2003, although these outbreaks are generally considered distinct from the outbreaks beginning in late 2003 and are not included in this total.
Adapted from World Organization for Animal Health, http://www.oie.int/.

sales value (in GBP) when looking at all Bernard Matthews brand products in the weeks preceding and following the avian flu outbreak announcement. Tesco, the largest supermarket chain in the United Kingdom, reported an immediate shift by shoppers from poultry to beef in the weekend following the announcement (Rigby and Wiggins, 2007).

While, overall, Bernard Matthews brand products saw a decline in sales value and sales volume, this trend was not shared by all turkey products. Table 12.7 shows changes in sales figures in the 3 weeks following the avian flu announcement indexed against the same figures for the 3 weeks preceding the announcement. While these figures show sales of Bernard Matthews brand Turkey Breast Roast declined immediately following the outbreak announcement, sales of Bernard Matthews brand Turkey Burgers were nearly four times higher than the pre-announcement level due to a buy-one-get-one-free price promotion.

Although Bernard Matthews brand Turkey Breast Roast saw an overall decline in sales following the outbreak announcement, this purchasing reaction was not demonstrated by all consumers. Figure 12.3 shows overall sales for Bernard Matthews brand Turkey Breast Roast during the same time periods as Table 12.7, but is displayed by life-stage segments. Older Adults and Pensioners appeared to have reduced their consumption compared to the average British consumer, whereas Families and Young Adults purchased more of the Turkey Breast Roasts than before the outbreak announcement.

In addition to heterogeneity of consumer response reflected in demographic segments, consumer purchasing behavior also appears to vary

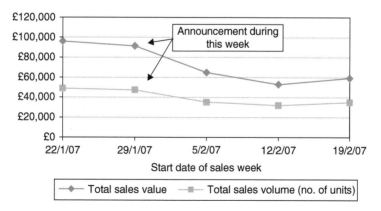

Fig. 12.2. Weekly sales volume and sales value of Bernard Matthews brand products at the time of the February 2007 avian flu outbreak.

Table 12.7. Differential impact on individual Bernard Matthews products indexed over time comparing periods of 8–28 January versus 29 January–18 February 2007.

	Total sales value	Total sales volume	Average price per unit	Total customers	Spend per visit	Units per visit
Turkey Burgers 283 g	251	452	55	306	89	160
Turkey Breast Roast 567 g	76	76	100	75	102	102

Adapted from Cacciolotti *et al.* (2007).
Index = 100.

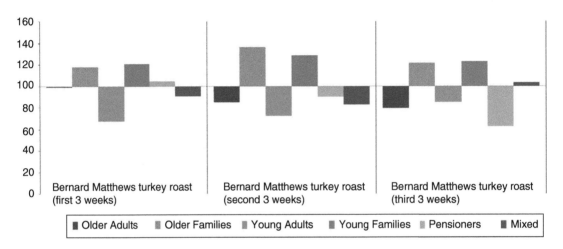

Fig. 12.3. Differential impact by consumer segment indexed against average British consumer. Adapted from Cacciolotti *et al.* (2007).

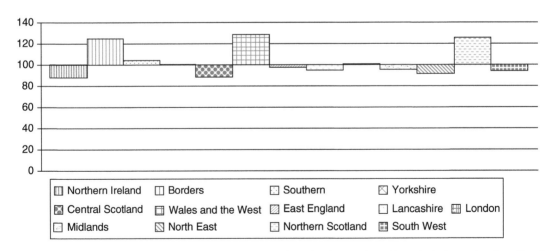

Fig. 12.4. Regional profile of purchases of Bernard Matthews brand turkey products 3 weeks before avian flu outbreak announcement indexed against the average British consumer.

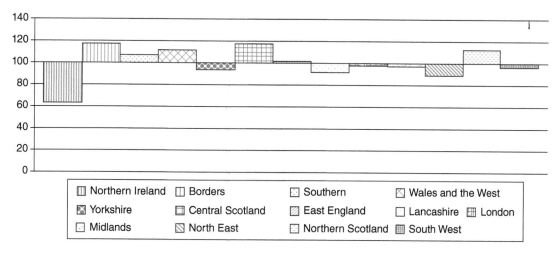

▥ Northern Ireland	▯ Borders	▯ Southern	▨ Wales and the West
▨ Yorkshire	▯ Central Scotland	▨ East England	▯ Lancashire ▥ London
▯ Midlands	▨ North East	▯ Northern Scotland	▥ South West

Fig. 12.5. Regional profile of purchases of Bernard Matthews brand turkey products 3 weeks following avian flu outbreak announcement indexed against the average British consumer.

according to the region of the country where the products were purchased. In Figs 12.4 and 12.5, regional purchasing profiles are indexed against the average British consumer for the periods of 8–28 January and 29 January–18 February 2007.

In the five regions where the purchasing of Bernard Matthews brand products was above the national average before the outbreak, purchasing continued to be above average in all regions except London, where it fell slightly below the index. It may be the case that familiarity with the brand allowed shoppers in these areas to continue to purchase Bernard Matthews brand products with more confidence and less perceived risk than consumers in other regions.

For the regions where consumers were purchasing at a level below the national average before the announcement, the only region to raise its purchasing behavior above the index was the East England region where the affected farm was located. In this case, familiarity with the industry and sensitivity to the economic impact of the food scare may have resulted in a reduced and arguably more realistic perception of the risk associated with Bernard Matthews brand turkey products available for purchase during this time.

Costs Associated with Ineffective Policy Design

Ineffective policy design could increase cost and harm related to foodborne pathogens. Safety and health policies are adopted to reduce harm to potential victims from accidents and other harmful events. However, such policies can induce offsetting behavior in which potential victims respond to the policies by relaxing their guard and increasing their exposure to the risk (Miljkovic *et al.*, 2009).

In other words, risk-reducing policies may affect people's risk preferences and, more specifically, cause them to take on more risks. Figure 12.6 presents the potential relationship between perceptions of risk, the actual degree of hazard, information, and demand for the risky commodity that can lead to this phenomenon. In communication theory, the gap between a person's perception of a risk and the actual degree of hazard can be represented by outrage (Sandman, 1987), which, being largely a product of fear of the unknown, is driven primarily by information and can significantly affect demand. Initially, lack of accurate information about food safety or an outbreak will move perceived risk from the baseline risk perception (α). Positive information via creation of mandatory food safety regulations can subsequently modify consumers' risk perceptions, bringing them gradually back to the baseline level (Lui *et al.*, 1998). Positive information also can provide consumers with a choice regarding whether to demand food that presents a relatively high risk, such as burgers served rare.

This is possible because the media reports on outbreaks of foodborne illnesses and recalls affect consumers' attitudes toward the risk and demand for hazardous products (Piggott and Marsh, 2004).

Food safety policies and offsetting behavior

In 1996, the USDA FSIS introduced mandatory PR/HACCPs following repeated discoveries of *E. coli* and *Salmonella* in the United States food supply in the 1980s and early 1990s. The new regulations required food processing plants to identify critical control points (CCPs) in their production and processing operations. These points are defined as "any step in which hazards can be prevented, eliminated, or reduced to acceptable levels. CCPs are usually practices/procedures that, when not done correctly, are the leading causes of foodborne illness outbreaks. Examples of CCPs include: cooking, cooling, reheating, holding" (University of Rhode Island, 2016). When consumers buy, process, or consume meat products, their perceptions of food safety depend extensively on perceived regulation of the entire production process. Consequently, consumers often view PR/HACCPs as more extensive than they actually are, creating imperfect information and mistaken reliance on positive food safety information from the media and/or federal agencies.

As shown in Fig. 12.7, outbreaks of foodborne illnesses peaked in 2000, a year in which numerous reports of positive information came from the media and some federal agencies about mandatory PR/HACCP implementation for all meat and poultry facilities. Nationwide, the number of (single-state and multi-state outbreaks combined) foodborne illness outbreaks per year almost tripled following

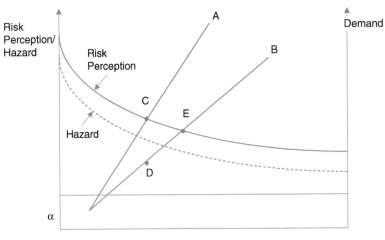

Fig. 12.6. Interaction between perceived risk, hazard, positive information, and demand.

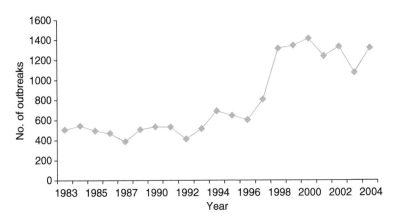

Fig.12.7. Number of outbreaks in the United States foodborne-disease outbreak surveillance system (Centers for Disease Control, www.cdc.gov).

implementation of PR/HACCP. During the same period, the number of meat processing facilities in the United States increased by less than 10% (US Census Bureau, 2002). The issue of imperfect information is magnified by the fact that different agencies are responsible for food safety at various points in the supply chain. For example, FSIS regulates most meat and poultry products and some egg products, while FDA is responsible for ensuring the safety of all other domestic and imported food products (Johnson, 2014).

It is important to understand the relevance of offsetting behavior in food safety applications and conditions under which policy changes affect consumers' risk behaviors. For instance, it is known that ground beef is more likely to contain pathogens such as *E. coli* than unground cuts such as steaks. Yet beef burgers are among the most popular foods in America. The safest way to prepare ground beef requires heating the beef patty until the center exceeds 71°C (160°F) (Food Marketing Institute and American Meat Institute, 1996).

Miljkovic *et al*. (2009) noted that consumers could be overconfident when they receive positive food safety information, causing them to drop their guard (offsetting behavior) and decreasing their consumption of well-done meat because the information they receive about improved food safety leads them to assume that the meat is safe. This allows them to consume it rare when that is their preference in terms of taste. Consumers may not be aware that the safety measures apply primarily to the processing stage and that contamination can occur at other points between processing and consumption and will not be eliminated in meat that is not thoroughly cooked.

Summary

The costs of foodborne illness extend beyond individual illness and deaths. They affect firms and society as a whole. To understand this, a SAM framework was developed by the USDA-ERS for HACCP. The reduction in losses of benefits from reduced illness and deaths was felt at all levels in society. The use of HACCP SAM also shows that appropriate mitigation strategies to reduce illness and deaths will generate more benefit to society. The SAM model analyzes the economic transactions and traces them for an individual firm or industry, the government, households, and the rest of the economy, which makes it easier to identify if a mitigation strategy like HACCP is beneficial. The level and distribution of the costs and benefits of the HACCP regulatory program for meat and poultry changed dramatically with economy-wide effects. SAM analysis provides policy makers with useful information about who ultimately benefits from reduced foodborne illnesses and who ultimately pays the costs of food safety regulations.

Costs related to food scares and infective policy design can be significant and should be included in the cost of foodborne pathogens to society.

References

Batz, M.B., Hoffman, S.A., and Morris, J.G. (2014) Disease-outcome trees, EQ-5D scores, and estimated annual losses of quality-adjusted life years (QALYS) due to 14 foodborne pathogens in the United States. *Foodborne Pathogens and Disease* 11, 395–402.

BBC News (2008) Timeline: Bird flu in the UK. Available at: http://news.bbc.co.uk/2/hi/uk_news/4882824.stm (accessed 30 December 2015).

Cacciolotti, L.A., Fearne, A., and Yawson, D. (2007) How do consumers respond to food scares? A case study of avian influenza in the UK. Using supermarket loyalty card data. *AAEA Annual Meeting July 29 to August 1*.

Doering, C. (2007) World overreacted to U.S. mad cow discovery: industry. *Reuters*, 15 November.

Food Marketing Institute and American Meat Institute (1996) *A Consumer Guide to Safe Handling and Preparation of Ground Meat and Ground Poultry*. Publication No. 458-016. Food Marketing Institute and American Meat Institute, Washington, DC.

FSA (2007) Food Standards Agency Annual Report 2006/07. Available at: http://www.food.gov.uk/multimedia/pdfs/annualreport200607.pdf (accessed 30 December 2015).

FSIS (2007) Food recalls/production and inspection/fact sheet. Available at: http://www.fsis.usda.gov/Fact_Sheets/FSIS_Food_Recalls/index.asp (accessed 30 December 2015).

Golan, E.H., Vogel, S.J., Frenzen, P.D., and Ralston, K.L. (2000) Tracing the costs and benefits of improvements in food safety. *United States Department of Agricultural Economics Report No. 791*. USDA, Washington, DC.

Hoffmann, S., Batz, M., and Morris, J.G. (2012) Annual costs of illness and quality-adjusted life year losses in the United States due to 14 foodborne pathogens. *Journal of Food Protection* 75, 1291–1302.

Huang, C.L. (1993) Simultaneous-equation model for estimating consumer risk perceptions, attitudes, and willingess-to-pay for residue-free produce. *Journal of Consumer Affairs* 27, 377–388.

Jin, H.J., Sun, C., and Koo, W.W. (2003) The effect of food-safety related information on consumer preference: the case of BSE outbreak in Japan, No. (506).

Center for Agricultural Policy and Trade Studies, North Dakota State University, ND, USA.

Johnson, R. (2014) The federal food safety system: a primer. Available at: https://www.fas.org/sgp/crs/misc/RS22600.pdf (accessed 30 December 2015).

Knowles, T., Moody, R., and McEachern, M. (2007) European food scares and their impact on EU food policy. *British Food Journal* 109, 43–67.

Lee, E., Hu, M.Y., and Toh, R.S. (2000) Are consumer survey results distorted? Systematic impact of behavioral frequency and duration on survey response errors. *Journal of Marketing Research* 37, 125–132.

Lui, S., Huang, J.C., and Brown, G.L. (1998) Information and risk perception: a dynamic adjustment process. *Risk Analysis* 18, 689–699.

Mark, D. and Halstead, M. (2007) Beef export verification programs: what should cattle producers do? Available at: http://www.ianrpubs.unl.edu/epublic/pages/publicationD.jsp?publicationId=576 (accessed 30 December 2015).

Miljkovic, D., Nganje, W., and Onyango, B. (2009) Offsetting behavior and the benefits of food safety regulation. *Journal of Food Safety* 29, 49–58.

Moon, W., Rimal, A., and Balasubramanian, S. (2004) Willingness-to-accept and willingness-to-pay for GM and non-GM food: U.K. consumers. *Paper presented at the American Agricultural Economics Association Conference*, Denver, CO, USA.

Nganje, W. and Wolf, M. (2008) Food scares and perception of risks. In: AAEA Annual Meeting, 27–29 July, Orlando, FL, USA .

Olsen, S.J., MacKinon, L.C., Goulding, J.S., Bean, N.H., and Slutsker, L. (2000) Surveillance for food borne disease outbreaks – United States, 1993–1997. *MMWR Surveillance Summaries* 49(SS01), 1–51.

Pennings, J.M.E., Wansink, B., and Meulenberg, M.T.G. (2002) A note on modeling consumer reactions to a crisis: the case of the mad cow disease. *International Journal of Research in Marketing* 19, 91–100.

Piggott, N.E. and Marsh, T.L. (2004) Does food safety information impact U.S. meat demand? *American Journal of Agricultural Economics* 86, 154–174.

Richards, T.J. and Patterson, P.M. (1999) The economic value of public relations expenditures: food safety and the strawberry case. *Journal of Agricultural and Resource Economics* 24, 440–462.

Rigby, E. and Wiggins, J. (2007) Poultry sales begin to fall. *Financial Times*, 10 February. Available at: http://www.ft.com/cms/s/0/35eb7360-b8ac-11db-be2e-0000779e2340.html?ft_site=falcon&desktop=true#axzz4G5xQ9E6R (accessed 30 December 2015).

Sandman, P.M. (1987) Risk communication: facing public outrage. *EPA Journal* 13, 21–22.

Schroeder, T.C., Tonsor, G.T., Pennings, J.M.E., and Mintert, J. (2007) The role of consumer risk perceptions and attitudes in cross cultural beef consumption changes. In: *Western Agricultural Economics Association Annual Meeting*, 29 July–1 August, Portland, OR, USA.

Smed, S. and Jensen, J.D. (2005) Food safety information and food demand. *British Food Journal* 107, 173–186.

Smith, C. (2007) When bird flu met brand. *Marketing* 7 February 23.

Smith, M.E., van Ravensway, E.O., and Thompson, S.R. (1988) Sales loss determination in food contamination incidents: an application to milk bans in Hawaii. *American Journal of Agricultural Economics* 70, 513–520.

Thomsen, M.R. and McKenzie, A.M. (2001) Market incentives for safe foods: an examination of shareholder losses from meat and poultry recalls. *American Journal of Agricultural Economics* 82, 526–538.

University of Rhode Island (2016) Hazard analysis of critical control points principles. University of Rhode Island Food Safety Education. Available at: http://web.uri.edu/foodsafety/hazard-analysis-of-critical-control-points-principles/ (accessed 17 October 2016).

US Census Bureau (2002) Rendering and meat byproduct processing: 2002. Available at: https://www.census.gov/prod/ec02/ec0231i311613.pdf (accessed 17 October 2016).

Verbeke, W. (2001) Beliefs, attitude and behavior towards fresh meat revisited after the Belgian dioxin crisis. *Food Quality and Preference* 12, 489–498.

Verbeke, W. and Viaene, J. (1999) Beliefs, attitude and behavior towards fresh meat consumption in Belgium: empirical evidence from a consumer survey. *Food Quality and Preference* 10, 437–445.

Verbeke, W., Ward, R.W., and Viaene, J. (2000) Exploring influencing factors on meat consumption decisions through probit analysis: the case of fresh meat demand in Belgium. In: *American Agricultural Economics Association Annual Meeting*, 8–11 August, Nashville, TN, USA.

Verbeke, W., Frewer, L.J., Scholderer, J., and De Brabander, H.F. (2007) Why consumers behave as they do with respect to food safety and risk information. *Analytica Chimica Acta* 586, 2–7.

Wansink, B. (2004) Consumer reactions for food safety crises. *Advances in Food and Nutrition Research* 48, 103–150.

13 Cost and Benefits of Control Measures: Food Traceability

> **Key Questions**
> - What is food traceability?
> - What technologies do we use to track the flow of food and food risks?
> - What is the role of traceability in mitigating the cost of food risks?
> - How does the cost of traceability compare to its benefits?

Introduction

In the current climate of foodborne illnesses from pathogens and increased food trade around the globe, a major concern involves the actions taken by the industry and governmental agencies to contain or trace the source of current and future outbreaks in a timely, targeted, and cost-effective manner. However, trace back and containment of food recalls often take several weeks or months, resulting in greater economic loss to the industry and to regional, and national economies. The duration and cost of trace back for food imports may be even more problematic. Box 13.1 provides an example.

There is little question that rapid-response and targeted trace-back systems can minimize economic damage inflicted by food safety events by speeding up and narrowing product recalls. Faster recalls avoid additional cases of illness or death, and targeted recalls avoid false alarms on products that are safe. Despite this awareness and the proliferation of traceability standards and systems, investigators in food supply networks often find that, when faced with an unexpected failure, participants "scramble" to produce the required information, leading to information losses or errors (Charlier and Valceschini, 2008) that turn into delays. One reason for these problems is that traceability systems are not uniformly implemented by all participants. When traceability is voluntary, some participants may try to avoid or reduce the costs of implementing traceability

systems. This suggests that enforcing centralized or cooperative standards for a rapid response among participants in food supply networks is a critical step in reducing trace-back times and executing more targeted recalls.

In this chapter, we investigate the costs and benefits of implementing the 24-hour traceability rule by examining the development of the California Leafy Green Marketing Agreement (LGMA, 2007). The LGMA was officially formed in September of 2007 in response to the 2006 *E. coli* O157:H7 outbreak associated with bagged baby spinach produced on farms located in the Salinas Valley of central coastal California (Stuart *et al.*, 2006). Presently, the LGMA consists of 120 growers, distributors and processors that account for approximately 99% of the volume of leafy greens (14 types of leafy greens) produced in California. Our case study provides an in-depth review of the costs and benefits associated with the implementation of traceability systems in California leafy green production, distribution, and retailing, which is an example of a tightly coupled, linear supply network. Costs are estimated for two electronic traceability systems: electronic barcode and radio frequency identification (RFID). The profit model developed by Pouliot and Sumner (2008) and data from the 2006 spinach outbreak are used to assess the cost-effectiveness of the trace-back response rate of the LGMA system.

The contribution of this case study stems from the analysis of the costs and benefits of implementing the 24-hour traceability rule. Specifically, we show that investments in rapid-response, targeted trace-back systems are likely to be cost-effective due to the minimized trace-back response rates. We also examine the major role of a tightly coupled, linear supply network, discussed in Chapter 1, when trying to implement traceability systems.

Traceability and Supply Network Structure

A complex network structure is one potential reason why firms tend to delay investments in rapid-response trace-back systems. Although traceability in the food supply network is often regarded as a product-level phenomenon, it is at least partially a function of the characteristics of the supply network structure (Roth *et al.*, 2008). In the LGMA, we develop the idea that traceability depends on characteristics of food supply networks that enable users to accurately reconstruct the chronology and flow of the steps in the production, distribution, and retail of the network's products.

The California leafy greens supply network is an ideal setting for a case study of this type, because its structure is likely to facilitate implementation of traceability systems. Skilton and Robinson (2009), adapting Perrow's (1999) theory of normal accidents to supply networks, proposed the leafy greens network as an example of a tightly coupled, linear supply system.

Perrow (1999) defined systems network complexity between: (i) complex systems and linear ones; and (ii) tightly coupled and loosely coupled systems (see Figure 1.6 in Chapter 1). He argued that system complexity is important because complex interactions multiply the opportunities for hard-to-understand accidents to occur. Complex systems have unplanned, unexpected, invisible, unfamiliar, ambiguous, or incomprehensible sequences, often made more obscure by poorly understood transformation processes. Less complex systems are likely to be more transparent because interactions will be less complex, transformations will be fewer, and the number and variety of components or actors involved will be less. As a result, linear systems afford a greater opportunity to get a clear picture of the whole system, making them easier to trace back.

On the other dimension, loosely coupled systems allow for processing delays, do not have fixed sequences or relationships, retain slack resources, and exploit fortuitous substitution possibilities (Skilton and Robinson, 2009). In tightly coupled systems, performance standards are enforced and unambiguous (e.g. marketing contracts). Delays are minimized, sequences are invariant, methods are constrained, any buffers or redundancies are designed, and substitutions or commingling are limited. Loosely coupled systems are less likely to be transparent because sensitive information is more likely to be ambiguous and less likely to be visible to and understood by the various actors in the system. Tightly coupled systems thus facilitate trace back by ensuring that information is captured, maintained, and readily available for rapid recall.

We argue that, compared to other supply networks, the California leafy greens industry exhibits the characteristics of a linear, tightly coupled supply network (see Box 13.2).

Changing either coupling or complexity alters traceability. A loosely coupled, linear supply network would also consist of mostly intransitive triads, without the constraint of highly specified relationships. In a network of this type, the number of paths through the network is relatively small, but, in the absence of tight coupling, information and material flows are less likely to retain their integrity. Examples of this type of network can be found in the supply of traditional and locally produced foods. There are typically only a few types of actor in these

In the typical leafy greens network, as shown in Fig. 13.1, multiple growers supply a packer, who ships the product either to a distribution center or to a re-packer. If the product is re-packed, it is shipped to a distribution center. Distribution centers, which can be controlled by third parties or by retailers, ship to retailers. The network is linearly structured, as can be seen by the limited number of transitive triadic relationships in the figure. A triadic relationship is transitive if A is connected to B, B is connected to C, and C is connected to A. In this case the only triadic relationships are between the packer, re-packer, and distributors, and between the packer, distribution center, and retailer distribution center. The entities depicted in the figure (such as growers' fields, packing facilities, distributors, and retailers) are arranged in chains that are only sparsely linked to each other. Within these chains, relationships are governed by rigorously enforced contracts that specify detailed information flows. In a network of this type, the number of paths through the network is relatively small. Because information flows and material flows are tightly linked, data are more likely to retain their integrity.

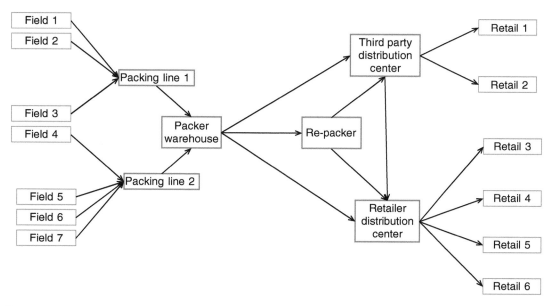

Fig. 13.1. A typical leafy greens supply network.

systems, which interact without contracts and without close supervision by regulators. Growers, processors, and consumers are local and small. While coupling is loose, it should be noted that the simplicity of the network can make trace back easy.

Trace back is more difficult when supply networks are complex. A tightly coupled complex supply network would be made up of many types of actor, arranged in networks in which connections are relatively dense and connected by relationships that are governed by rigorously enforced contracts specifying detailed information flows. Skilton and Robinson (2009) suggest that networks of this type may be rare because of the competing pressures of network complexity and tight coupling. Complexity tends to make data integrity difficult to achieve, particularly when transformations and commingling are common. Trying to manage this by imposing tight coupling is likely to result in reductions in complexity.

This suggests that loosely coupled complex supply networks will be more common than tightly coupled complex ones. Many food supply networks have this form, which is the type in which traceability is most difficult. An example of this type of network is a broker-based network supplying commodity inputs, particularly those that have been transformed, such as ground beef. While it is in these complex supply networks that trace back is most difficult (Moss, 2009), the fact that traceability is also a problem in tightly coupled linear ones leads us to try to understand the simpler case, represented by the LGMA, first. We make this attempt in the hope of arriving at a better understanding of the costs and benefits of system-wide implementation that will be at least in part transferable to more complex cases.

Case study: Traceability systems within the LGMA

The LGMA does not require members to implement specific systems. Although it encourages the use of tracking systems based on RFID, no specific technology or information system is mandated by the agreement itself. Instead, the agreement makes reference to federal law, the only explicit specification for compliance with the agreement. The LGMA audit checklists' specific requirements are given as follows (LGMA, 2008):

1. Is an up-to-date growers list with contact and location information available for review?
2. Is the handler in compliance with the registration requirement of The Public Health Security and Bioterrorism Preparedness and Response Act of 2002?
3. Does the handler have a traceability process?
   a. Does it enable identification of the immediate non-transporter source?
   b. Does it enable identification of the immediate non-transporter subsequent recipient?

While the implementation of a specific electronic traceability system is not mandatory in the LGMA requirements, the majority of members have deployed barcode or RFID technologies to meet the requirements of the marketing agreement (Estrada-Flores, 2009). Barcodes are still the predominant technology used by leafy green growers, distributors, and retailers, but because they require a line of sight, barcode data are often captured using handheld devices. Handheld devices increase the probability of human error and require batch uploading, increasing the difficulty of ensuring data integrity. Cost limitations have been cited as a major reason why other real-time technologies are not extensively used. RFID can enhance the rapid response by improving and automating capture of logistics data, permitting

near real-time tracking, which in turn enables logistics managers to proactively respond to events in the supply chain, such as breaches of the cold chain (temperature abuse).

Regardless of the specific technology deployed, identifying tags are placed on boxes, pallets, and/or the product packaging, in which the assigned unique number provides the information necessary to trace the product back to the immediately preceding stage of production. Tag data frequently record, but are not limited to, grower, ranch location, planting block/lot, planting date, harvesting date, harvesting crew, ship date, ship-to locations, manufacturing plant, production shift and line, production date, and a "Best if Used By" date (Church Brothers Produce, 2009; Estrada-Flores, 2009; Sunridge Farms, 2009).

As depicted by the audit requirements of the LGMA and by the description of a number of LGMA members' traceability systems, the success of a traceability system relies heavily on comprehensive documentation and record-keeping procedures. For example, Growers Express' field managers maintain records regarding staff and all products and materials used during the production process. These records are primarily electronically stored. Growers Express maintains that pertinent trace-back data can be made available for review by investigators within 2 hours of notice (Growers Express, 2009). The tightly coupled linear network of LGMA participants involves as many as five stages that link the retailer to the farm (see Fig. 13.1). These linkages represent both information flows and the physical distribution and storage of food shipments (Estrada-Flores, 2009). See Box 13.3 for examples.

A survey conducted by Tootelian (2008) revealed that, prior to the 2006 *E. coli* California spinach outbreak, 89.8% of all growers indicated they had traceability in place. It should be noted that 100% of growers shipping more than 1,000,000 lbs had traceability systems. However, the predominant traceability systems were a paper trail and barcode technology, making it challenging to conclude that these systems were able to trace back within 24 hours for each participant involved. The survey indicated that, after the 2006 outbreak, 60.5% of growers indicated they had expanded their traceability programs to more electronic-form systems, including real-time tracking.

Because the kinds of technology deployed can alter the ability to achieve 24-hour trace back, in the next section we provide traceability cost estimates

The following two examples provide insight on the trace-back rate for current electronic users. In August 2007, Metz Fresh, LLC employed their traceability system after finding that 8,000 cartons of fresh spinach tested positive for *Salmonella*. Within 3 days of harvest, stores and restaurants were notified of the product recall, whereby more than 90% of the tainted spinach never reached the market (CIDRAP, 2007). In the summer of 2009, Tanimura & Antle Inc. recalled romaine lettuce after random testing conducted by the Wisconsin Department of Agriculture found traces of *Salmonella*. Tanimura & Antle Inc. was informed of the possible contamination on 20 July, which was the same day that the product was distributed to 29 states, Canada, and Puerto Rico (Withers, 2009). Tanimura & Antle Inc. was able to identify that the potentially contaminated lot (code 531380) was harvested between 25 June and 2 July and alerted their customers of the recalled product within hours of being notified (FDA, 2009).

for two trace-back technologies currently or potentially used by Leafy Green members; barcodes (which are the most commonly deployed base technology) and RFID (which we use to provide an upper cost range).

Cost of the LGMA traceability system

Adhering to LGMA standards and meeting the audit compliance requirements can be costly for members. Paggi (2008) noted that LGMA compliance costs range between US$210 and US$260 per acre. Of this cost, approximately US$50 per acre is estimated to be record-keeping costs. Annual investment in food safety for LGMA members has almost tripled since the introduction of LGMA. According to the 2007 LGMA status report, total LGMA annual food safety investments prior to September 2006 were US$23.7 million, compared to US$71.3 million after September 2006 (Tootelian, 2008). This translates to an average annual investment of US$604,545 per member enterprise after September 2006. Combining the estimated annual operating costs for LGMA audit compliance (food safety employee costs, annual water testing expenses, annual LGMA membership funding), and the total estimated annual investment for LGMA compliance, the estimated total costs for LGMA members' compliance range from approximately US$80 to US$91 million per year. These costs would cover expenses incurred by growers and handlers, and the traceability systems in this tightly coupled, linear network system could represent costs for tracing production forward from producer to distributor. This estimate implies that average annual expenses relating to compliance to the LGMA range from US$0.0128 to US$0.0158 per pound (Table 13.1).

We estimate that a significant percentage of the food safety cost (20–45% of the total LGMA compliance cost) is attributed to record keeping and traceability. However, the actual cost share varies widely for individual enterprises based on the type of traceability system used (paper records, barcodes, and RFID).

Costs for two technologies (i.e. barcode and RFID) were estimated for representative firms by size, and then aggregated to obtain the estimated industry costs. Each system's costs, which included both variable and fixed costs, were computed for representative firms of three sizes measured in shipment volume (1 = 0–100,000 lbs; 2 = 100,001–999,999 lbs; 3 = 1 million lbs or more). Among the actual LGMA membership (n=118), 34.3% were in size category 1, 36.3% in size category 2, and 29.4% in category 3 (Tootelian, 2008). The related costs for firms in each of the three size categories were aggregated by their share of the industry members to get the total industry costs. Industry costs and parameters were based on Tootelian's 2008 report, state/industry statistics, and other published documents. The cost estimates were estimated from the volume of leafy greens of the California leafy greens industry, which represents approximately 75% of the total United States volume (LGMA, 2008).

For each of the technologies, total fixed cost is the sum of the individual fixed cost components. The fixed costs were depreciated over 5 years using a discount rate of 10% as the discount value for the cost of working capital. Variable costs differ by the size of the member groups. Total variable costs include the sum of the variable costs based on their assignment to the traceability system. For example, a barcode-based system has

---

Table 13.1. Estimated annual industry costs relating to LGMA compliance.

FOOD SAFETY EMPLOYEE WAGES	
US average salary for food safety inspectors	US$37,599
Approximate number of food safety employees	267
TOTAL ANNUAL FOOD SAFETY	
Employee wages	US$10,038,933
IRRIGATION WATER TESTING EXPENSES	
Average cost for water testing (per test)	US$42–US$70
Approximate number of annual water tests	73,956
TOTAL ANNUAL WATER TESTING	
Expenses	US$3,106,152–US$5,176,920
Total annual LGMA membership funding	US$4,500,000
Total estimated annual operating costs (LGMA audit compliance)	US$17,645085–US$19,715,853
Total estimated annual investment expenses	US$62,092,780–US$71,000,000
Total estimated costs relating to LGMA membership	**US$79,737,865–US$90,715,853**
Approximate annual volume of leafy greens (lbs)	5,720,000,000–6,240,000,000
(approx. 22–24 lbs per carton)	
Total estimated annual costs (LGMA audit compliance) per lb	**US$0.0128–US$0.0158**
Total estimated annual costs attributed to traceability per lb (20–45%)	**US$0.00256–US$0.00711**

Source: Compiled from Cline (2007), Tootelian (2008) and USDA (2009).

variable costs for the three size groups of US$1,868,819 (sum of US$91,405, US$192,281, and US$1,585,133). These costs are simulated from the volume of shipments for all members in that size category. The variable costs include costs for the barcode labels, barcode label printer, barcode hand-held reader, and employee training. The total fixed cost for the barcode system is US$1,393,258. Total cost is the sum of fixed and variable costs. For the barcode system, the total cost is US$3,262,077, and applies if all firms adopted the barcode system.

The total industry costs were estimated for each of the two technologies (i.e. RFID and barcode) by aggregating the costs incurred by the firms in the industry across the three sizes of firms. For both technologies, total fixed and variable costs were highest for the RFID system (US$109 million for passive tags and US$1,372 million for active tags). As shown in Table 13.2, costs also vary by firm size or sales volume (firm size categories assigned to member type 1, 2, or 3). For example, the RFID variable costs range for small and large sales volume varied between US$132,905 and US$87.5 million for passive tags. The variable costs for the barcode system ranged from US$91,000 to US$1.6 million for the small and large sales volume, respectively.

In estimating cost and technology configurations for traceability systems, it should be noted that, when commingling is prevalent (for loosely coupled, linear and complex systems), multiple barcodes would be needed. Barcode technology was determined to be the least expensive in a tightly coupled, linear system, such as spinach. However, the usage of active RFID tags and technologies that enable data to be rewritten at multiple locations could become cost-effective when a requirement for multiple barcodes exists. Hence, requiring and enforcing information one step back and one forward has implications in this example for both cost and the technology selected depending on the firm's size and the processes done by each enterprise.

Analysis of costs–benefits of the LGMA traceability system

LGMA, consisting of growers, distributors, and processors, represents an appropriate setting to study the costs and benefits related to the implementation of traceability systems in the supply chain network. LGMA is particularly relevant to the examination of food supply chains as traceability depends on various features of supply networks that enable users to reconstruct, as accurately as possible, the chronology of the steps in the production, distribution, and retail of the network's products.

The 2006 California spinach *E. coli* outbreak is mainly relevant for the analysis of the supply chains' losses and the costs associated with fresh product recalls. This is especially the case considering that each partner of the supply chain is likely to be negatively affected by an outbreak. For example,

Table 13.2. Annual costs associated with the implementation of a traceability system.

Costs associated with RFID system	Estimated costs
Total fixed costs[a]	US$21,168,106
Total variable costs – passive tag	
Total variable costs (Member 1[b])	US$132,905
Total variable costs (Member 2[c])	US$518,006
Total variable costs (Member 3[d])	US$87,450,419
Total variable costs – active tag	
Total variable costs (Member 1[b])	US$536,238
Total variable costs (Member 2[c])	US$4,853,840
Total variable costs (Member 3[d])	US$1,345,652,954
Costs associated with barcode system	**Estimated costs**
Total fixed costs[a]	US$1,393,258
Total variable costs	
Total variable costs (Member 1[b])	US$91,405
Total variable costs (Member 2[c])	US$192,281
Total variable costs (Member 3[d])	US$1,585,133

[a]Depreciated over 5 years.
[b,c,d]The costs were computed for representative firms of three sizes measured in shipment volume (1 = 0–100,000 lbs; 2 = 100,001–999,999 lbs; 3 = 1 million or more lbs).

the total retail value of Dole baby spinach recalled was US$163,380. Furthermore, the total lost productivity expenses due to *E. coli* O157:H7 reached US$381,879.84.

Due to the problems associated with traceability, the contaminated spinach was traced to National Selected Foods as the packer; however, the farm of origin could not be identified with the same rapidity (Pouliot and Sumner, 2008). Consequently, proposals of better management practices along the supply chain have been triggered by this outbreak.

The estimated benefits of having a traceability system in place will be reflected in the reduction of costs due to contamination. We use the 2006 *E. coli* O157:H7 outbreak in our analysis because its severity continues to be startling; specifically, 204 persons were infected, including 31 cases of hemolytic–uremic syndrome (a serious complication that can cause kidney failure) and three deaths. Moreover, in the decade preceding the 2006 epidemic, nine other *E. coli* O157:H7 outbreaks were linked to lettuce or spinach grown in the Salinas Valley region of California (Cooley *et al.*, 2009). The personal injury and lives lost continue to be the utmost regrettable consequence from this outbreak, yet the economic and financial fallout also severely impacted the public. The entire United States spinach industry experienced financial losses and reputation damage,

while consumers' confidence in food safety diminished and public funding was allotted to recoup the costs of the outbreak.

We estimate the failure costs associated with the 2006 *E. coli* O157:H7 outbreak based on the United States spinach industry losses and the total costs linked to the product recall to be approximately US$129 million. The industry cost of the outbreak was estimated at US$80 million, but some studies have reported industry loses in the range of US$100 million to over US$350 million (McKinley, 2006; Weise and Schmit, 2007).

The total volume of contaminated product that caused the outbreak and the product recalls were approximately 15,750 lbs of bagged Dole baby spinach, which were identified by the code P227A. The code indicated that the spinach was produced at Natural Selection's south plant (P) on the 227th day of the year (15 August) during the first of two shifts (A) (Weise and Schmit, 2007). Although several leafy green products were recalled due to the outbreak, the 15,750 lbs of bagged Dole baby spinach was deemed responsible for the outbreak and product recalls. Losses due to false alarms for commingled products or spinach produced out of California could have been avoided if trace back had been rapid (occurred within 24 hours for each segment in Figure 13.1), and technology could lead to a more targeted recall (see Box 13.4).

Box 13.4. Assumptions to estimate benefits

In order to simulate the traceability benefits and cost-effectiveness of the LGMA systems, we assume that trace-forward response rates could improve significantly from 47 days to less than 50% (24 days) and 75% (12 days) with rapid response, targeted systems like RFID and GS1 systems. This assumption is realistic, as the survey by Tootelian (2008) revealed that about 60.5% of those who had paper trail or barcode traceability systems expanded their systems. This expansion should include better electronic tracking systems. The assumptions of increased use of electronic systems and availability of linking the electronic information across agents in the food system are consistent with the program recommended by the panel for 24-hour electronic data availability (IFT, 2009).

We simulated the benefits of cost reduction using the profit model developed by Pouliot and Sumner (2008) and the assumptions on trace-forward response rates above. Their model includes three main components for net profits; specifically, they calculate the profit by subtracting the costs of providing traceable and safe produce, and the costs associated with unsafe food delivery from the gross revenue. We implement this model and attempt to show the benefits that forward traceability provides by adding the costs due to spinach recalls. When an outbreak occurs, if one looks exclusively back from the consumer to producer (i.e. trace backward), traceability systems are likely to be always considered too costly to be worth the investment. However, when trace forward is engaged, the costs to place the produce on the market are avoided and thus benefits are obtained.

In our model, the gross revenue is the revenue obtained for each pound of spinach. For traceability cost, we use the triangular distribution for the total estimated annual costs attributed to traceability per pound. The triangular distribution was also used for the expected loss from the spinach outbreak. That is, we considered the cases when there are no benefits from traceability, when the costs due to the spinach outbreak are reduced by 50%, and when the costs are eliminated completely.

We conducted @Risk analysis to show how likely the gross revenue for each pound of spinach is to exceed US$1.297 and, likewise the probability that this revenue will drop below this number, given that we implement a forward, rapid, and targeted traceability system. Our simulation results show that, if traceability is implemented, the probability that the revenue will be the same is 13%. Most important, the results illustrate that the likelihood of having negative returns/net revenue is 0%, given that we implement traceability systems. Simulated benefits also suggest that rapid-response targeted systems could have mitigated between US$9,795,564 and US$93,562,205.

Although the range of improved benefits is large, the study indicates the significant improvement possible through having access to supplies and product destination information through electronic means. The estimated costs and simulation employed in the case of the LGMA for spinach indicates that there are significant savings from a more rapid response. This response results from information technologies that improve the ability to track product flow. Compliance costs of LGMA membership (of US$0.0128–US$0.0158 per pound, with the associated record keeping and traceability costs ranging from US$0.0026 to US$0.0071 per pound) are significantly lower than the potential benefits of avoiding the 2006 *E. coli* O157:H7 outbreak and product recall. The costs are lower than the benefits achieved when more rapid and targeted recall systems (24-hour system for each participant) reduce the trace-forward response time by 50% (24 days) or 75% (12 days) in this case. Some of these costs are expected to be passed forward to consumers. For firms, costs of having a system in place would be recurring costs to the industry. Any individual firm may not experience a recall within a year, as this is likely to be a relatively rare event. For the industry, having a rapid response system in place reduces the costs (provides benefits) associated with a specific recall whenever this occurs. However, private benefits are also likely to be achieved when more rapid and targeted recall systems are implemented. For example, when an outbreak occurs, consumers "no longer trust the competence and/or integrity of parties involved in ensuring the quality of particular food products" (Hennessy *et al.*, 2003). This can be translated into lower to no consumption of demand for that particular product, which is likely to cause considerable costs to the industry as well as to private businesses. Therefore, while public benefits or avoided costs might not trigger the incentives of implementing traceability systems, the private benefits for not losing customers should be given careful consideration.

From a different perspective, public benefits (avoided costs) are driven by the private benefits. That is, publicized lawsuits due to food safety litigations are associated with considerable liability costs and loss of reputation for enterprises, which represents a significant incentive for firms and farms to supply safer food (Pouliot and Sumner, 2008). As a consequence, the public costs of foodborne illnesses are reduced. In other words, the private avoided costs due to the implementation of traceability systems in the production, distribution, and retailing of fresh produce are likely to drive the public benefits.

Again, these are conservative benefit estimates but are appropriate since not all LGMA members currently have 24-hour rapid-response trace-back systems in place. It is true that the benefits could be even larger if recalls are made before products even enter the retail outlet or are sold to end consumers. These results reveal that, with advancements in policy and adoption of rapid-response trace-back technology, the industry's benefits may by far outweigh the costs. Net present value and dynamic analysis suggest even more savings as the cost of technology will continue to decline.

According to Pouliot and Sumner (2008), as the number of firms and marketers within the supply chain increases, "industry reputation incentives for individual firms to supply safe food decreases." Thus, according to the same authors, without available traceability, firms are likely to be anonymous and the safe food supply chain is characterized by a free-rider problem. This issue highlights a greater role of traceability for industry-avoided costs as opposed to those of individual firms. In other words, when the free-rider problem exists, the industry has more incentives to increase traceability in order to avoid the costs associated with an outbreak.

Food Policy Implications

The important policy debate in the United States and other countries is whether or not to mandate traceability for domestic and international food supply chains. In the United States, for example, the FDA is considering a 24-hour rapid-response trace-forward and -backward electronic system. The emphasis of such a program in prior studies has mostly been on the cost of implementation. We expand the discussion to show that, in the case of California LGMA, there are clear benefits to implementing rapid-response and targeted traceability systems in food supply networks. If recalls are not conducted in a timely and targeted manner, even participants whose product is not contaminated suffer economic losses.

Following Skilton and Robinson (2009), we proposed that traceability is partly a function of the characteristics of the food supply network, and system complexity could not be overlooked. The LGMA supply network structure is tightly coupled and linear, which we have argued is the most desirable structure for achieving traceability. Estimating the cost of traceability of such a system is less complicated and direct. The problem that is revealed by introducing network structure as a predictor of mandating traceability is that many food supply networks are either loosely coupled or complex. Both of these structural characteristics impede traceability. One of the most important implications of this theory is that traceability systems must be tailored to the supply networks they are deployed in. Tightly coupled linear networks present the fewest challenges to systems developers, so it is no surprise that effective traceability systems are already common in these networks. Formal systems are much less common in loosely coupled linear networks, in part because there are very few steps between the raw material and the final consumer. Consequently, the costs to small players of implementing formal systems are high relative to the benefits in terms of traceability.

As networks become more complex, traceability systems must follow suit. In complex networks, the challenge is how the traceability system would preserve information through transformative processes, commingling of shipments, and other information-degrading processes that characterize complex networks. Effective trace back/forward in complex tightly coupled systems may depend on the presence of large, powerful players such as global food companies or global retailers in the network. These participants have an interest in rapid trace back because they have investments in brands that are put at risk by adverse events. Strong central participants would have to create a climate of enforcement throughout the network. Wal-Mart has begun to take this approach as evidenced by the retailer's recently announced sustainability initiative (Rosenbloom, 2008).

This initiative will result in a network with fewer suppliers operating in more tightly coupled relationships. A key issue for firms seeking to influence network practices will be their "reach" or how far their influence extends into the network of suppliers. Since controlling a complex network requires information-intensive relationships with a greater variety of counterparties, there is a risk of information overload for the controlling firm. In tightly coupled networks, reducing complexity and assigning an enforcement role to a central player would make traceability systems possible, but only at a high cost in terms of system resources, data quality, and the opportunity costs of committing to a smaller supplier base for any product.

Summary

In a loosely coupled complex network, we think solutions are limited to very simple logistics approaches similar to the one deployed by the LGMA. The problem in these networks is the absence of a central player, which suggests that government intervention, perhaps on an international scale, will be necessary. It is also possible that solutions will emerge from technology suppliers. There is a global consortium designing a standard protocol for logistics traceability based on RFID tags (IBM, 2007), but the systems' cost continues to be limiting. Electronic barcode costs are a fraction of RFID costs, but they do not have the same capabilities for enabling a rapid response. However, advances in technologies will continue to drive the costs of RFID and other rapid-response, targeted technologies lower and more affordable over time.

Rapid-response, targeted technologies have significant long-run benefit potentials. Potential savings from a single outbreak like the 2006 E. coli O157:H7 indicate benefits of minimizing recall loss, which may by far outweigh the cost of compliance of the LGMA traceability investments (except in the implementation of active RFID tags for large growers in this case). Noticeably, the compliance costs of LGMA membership (US$0.0128 to US$0.0158 per pound, with the associated record keeping and traceability costs ranging from US$0.0026 to US$0.0071 per pound) are significantly lower than the potential benefits of avoiding the 2006 E. coli O157:H7 outbreak and product recall. Since the implementation of the LGMA and its traceability system, there have been several timely interventions and fewer recalls, mostly voluntary lower class recall. This study makes its primary contribution to the goal of greater traceability by trying to show how the conditions that enable or impede traceability emerge from supply network structure. The study also suggests that investments in rapid-response, targeted trace-back systems could be cost-effective by minimizing trace-back response rates.

References

Charlier, C. and Valceschini, E. (2008) Coordination for traceability in the food chain: a critical appraisal of European regulation. *European Journal of Law and Economics* 25, 1–15.

Church Brothers Produce (2009) *Church Brother's Summarized Product Recall Procedures*. Available at: http://trueleaffarms.com/Newsandevents/Recall-Procedures.pdf (accessed 30 December 2015).

CIDRAP (2007) Foodborne disease rates changed little in 2007. Center for Infectious Disease Research and Policy. Available at: http://www.cidrap.umn.edu/news-perspective/2008/04/foodborne-disease-rates-changed-little-2007 (accessed 30 December 2015).

Cline, H. (2007) Leafy green food safety program proving successful. *Western Farm Press*. Available at: http://westernfarmpress.com/mag/farming_leafy_green_food/ (accessed 30 December 2015).

Cooley, H., Christian-Smith, J., and Gleick, P. (2009). *Sustaining California Agriculture in an Uncertain Future*. Pacific Institute, Oakland, CA, USA.

Estrada-Flores, S. (2009) Emerging trends in US logistics part II: food safety trends. Available at: http://www.food-chain.com.au/Dispatch3.pdf (accessed 30 December 2015).

FDA (2009) Tanimura & Antle voluntarily recalls one lot of romaine lettuce because possible health risk. Available at: http://www.fda.gov/Safety/Recalls/ucm173185.htm (accessed 30 December 2015).

Growers Express (2009) Big on safety. Available at: http://dailyquote.growersexpress.com/news/0428009_ge_greengiantrights.html (accessed 30 December 2015).

Hennessy, D.A., Roosen, J., and Jensen, H.H. (2003) Systemic failure in the provision of safe food. *Food Policy* 28, 77–96.

IBM (2007) Enabling RFID traceability in EPCglobal networks, intelligent information systems. Available at: http://www.research.ibm.com/labs/almaden/index.shtml (accessed 30 December 2015).

IFT (2009) Traceability (product tracing) in food systems. Institute of Food Technologists/Food and Drug Administration Contract No. 223-04-2503. Available at: https://www.diagraph.com/Portals/0/IFT%20Task%20Order%206%20Traceability%20in%20

Food%20Systems%20Economic%20Report.pdf (accessed October 17, 2016).

LGMA (2007) Food Safety Practices. Available at: http://www.lgma.ca.gov/food-safety-program/food-safety-practices/ (accessed 16 January 2017).

LGMA (2008) Government audit. Available at: http://www.lgma.ca.gov/food-safety-program/government-audits/ (accessed 30 December 2016).

McKinley, J. (2006) Center of *E. coli* outbreak, center of anxiety. Available at: http://www.nytimes.com/2006/09/25/us/25ecoli.html (accessed 30 December 2015).

Moss, M. (2009) *E. coli* path shows flaws in ground beef inspection. *New York Times*, 9 October, A1.

Paggi, M. (2008) An assessment of food safety policies and programs for fruits and vegetables: food-borne illness prevention and food security. Presented at *North American Agrifood Integration Consortium Workshop V: New Generation of NAFTA Standards*, 22 May, Austin, TX, USA. Available at: http://naamic.tamu.edu/austin/paggi.pdf (accessed 30 December 2015).

Perrow, C. (1999) *Normal Accidents: Living with High Risk Technologies*. Princeton University Press, Princeton, NJ, USA.

Pouliot, S. and Sumner, D.A. (2008) Traceability, liability, and incentives for food safety and quality. *American Journal of Agricultural Economics* 90, 15–27.

Rosenbloom, S. (2008) Wal-Mart to toughen standards. *The New York Times*, 22 October.

Roth, A.V., Tsay, A.A., Pullman, M.E., and Gray, J.V. (2008) Unraveling the food supply chain: strategic insights from China and the 2007 recalls. *Journal of Supply Chain Management* 44, 22–40.

Skilton, P.F. and Robinson, J.H. (2009) Traceability and normal accident theory: how does supply network structure influence the identification of the causes of adverse events? *Journal of Supply Chain Management* 45, 40–53.

Stuart, D., Shennan, C., and Brown, M. (2006) Food safety versus environmental protection on the central California Coast: exploring the science behind an apparent conflict. The Center for Agroecology & Sustainable Food Systems. Available at: http://escholarship.org/uc/item/6f90g0dg (accessed 30 December 2015).

Sunridge Farms (2009) Good agricultural practices. Available at: https://www.sunridgefarms.com/news/ (accessed 30 December 2015).

Tootelian, D. (2008) California Leafy Green Products, 2007 Signatory Survey. Summary Report of Findings. Available at: http://www.perishablepundit.com/docs/leafygreens-summary-finalreport.pdf (accessed 16 January 2017).

USDA (2009) Justification of proposed federal marketing agreement for leafy green vegetables. Available at: https://www.ams.usda.gov/?dDocName=%2520STELPRDC5077207 (accessed 30 December 2015).

Weise, E. and Schmit, J. (2007) Spinach recall: 5 faces, 5 agonizing deaths. 1 year later. Available at: http://a.abcnews.com/Business/Story?id=3633374&page=1 (accessed 15 December 2016).

Withers, D. (2009) Updated: traceback system speeds romaine recall. Available at: http://www.thepacker.com/fruit-vegetable-news/updated_traceback_system_speeds_romaine_recall_122123104.html (accessed 30 December 2015).

14 Impacts on Global Trade and Regulations

> **Key Questions**
> - How do foodborne pathogens and other agents impact trade?
> - How do nations use standards and regulations to restore trust in our food supply system?
> - What are the similarities and differences among measures to restore trust in our global food supply?

Global Trade Impacts of Foodborne Pathogens

Consumer demand escalates food trade and requires delivering more tonnage through ports of entry (POEs), rail, trucks, and air. Increased volumes of food trade with ever increasing velocity have been associated with significant food safety risks (unintentional food contamination from pathogens, or chemical or physical agents) and food defense risks (intentional food contamination by disgruntled employees or terrorists). While import inspections should help protect against outbreaks of foodborne illnesses, it is neither possible nor optimal to inspect all produce at the POE. Only a very small percentage of trade is inspected, in some instances less than 1% (Schmidt, 2007).

With increased international trade on the marketing system and limited resources to inspect and test all products for pathogens, these pathogens could spread fairly rapidly across our borders causing significant economic loss. For example, human illness costs due to seven foodborne pathogens in the United States alone resulted in 3.3–12.3 million cases of foodborne illness and up to 3,900 deaths, costing an estimated US$6.5–US$34.9 billion annually (Buzby and Roberts, 1997). They noted that the presence of foodborne pathogens in a country's food supply not only affects the health of the local population, but also represents a potential for spread to visitors and consumers in these countries trading food products. Consequently, outbreaks could result in international trade rejections and trade bans by each country.

Increased food trade creates challenges to control food safety or food defense risks. In the last century, there were several documented cases where pathogenic agents were used to infect livestock or contaminate food intentionally. Ecoterrorist factions have used plant toxins in Africa (Carus, 1999), anthrax in the UK (Chalk, 2003), and potassium cyanide in Sri Lanka (Cameron et al., 2001) to intentionally contaminate food.

Many and probably most food contamination cases go unreported because the contamination sources are very difficult to identify. Even when people do not die from contaminated food, the economic loss and market failure impacts can be substantial. When limited inspection resources are not efficiently distributed, market failure may arise from negative externalities or the public good nature of food protection. Negative externality may occur when some participants in the supply chain implement a food protection measure but are impacted by a food recall due to others who have not implemented similar recommended measures. When inspection systems fail to mitigate outbreaks from credence-type food protection, then market failure can be attributed to the public good nature of food protection, as the public demands a minimum level of safety.

Challenges with increasing food trade

Food trade is increasing around the globe due to a growing middle class and several other factors. Regmi (2001) noted that higher income, urbanization, demographic shifts, improved transportation, and consumer perceptions regarding quality and safety are changing global food consumption patterns. These shifts in food consumption have led to increased trade. For example, American consumers continually demand more fresh produce and food throughout the year, in particular during non-productive United States seasons (Table 14.1). Consumer demand escalates food imports and trade, and requires delivering more tonnage through the global food supply network. Similar trends are observed with different countries.

The demand for Mexico-grown fruits and vegetables in the United States is increasing substantially because off-season demand is not being met by domestic production. Approximately 6.2 billion lbs of fresh fruits and vegetables were imported from Mexico to the United States in 2005, 6.49 billion lbs in 2006, and 7.24 billion lbs in 2007 (USDA-FAS, 2008). The largest share, approximately US$2 billion dollars, of Mexico-grown fresh produce is imported into the United States through the Nogales, Arizona, POE during the winter months.

Increased volumes of imported foods with ever increasing velocity have been associated with significant food safety and food defense risks. Historically, firms may have considered supply chain risks and defense in the context of the potential threats and disruptions to their own operations. However, the interconnectedness of firms, products, and transportation infrastructure in high-speed global supply chains multiplies the potential costs of these risks and may lead to complete market failure without adequate produce supply chain safety and defense.

One example of market disruption is the 2008 *Salmonella enterica* outbreak of fresh jalapeño and serrano peppers from Mexico, which caused at least 1,329 cases of salmonellosis food poisoning in 43 states throughout the United States and in the District of Columbia. Nationwide, about 257 people were hospitalized and two deaths were associated with the outbreak. A second contamination event was the outbreak of hepatitis A that occurred in Tennessee, North Carolina, Georgia, and Pennsylvania in 2003. In Pennsylvania, over 650 people were infected and four people died as a direct result of this outbreak. Public health officials used genetic sequencing techniques to trace the outbreak back directly to green onions grown on farms in Mexico. These outbreaks cause significant public health and market disruption problems that may result in complete market failure and trade restrictions. Foods can also be exposed to intentional adulteration with biological, chemical, physical or radiological agents by a terrorist, which is called agroterrorism.

Current Standards and Regulations

Mitigating the impacts of food risks on trade involves private–public partnership. Countries develop policies and the private sector develops certifications and standards to mitigate trade impacts from food risks and restore consumer confidence on safe trade flows. Several policies have been developed and implemented in the United States. Examples of public sector initiatives include pathogen reduction (PR)/hazard analysis and critical control points (HACCP), the 2004 Bioterrorism Regulation, and the 2010 Food Safety Modernization Act. Examples of private sector initiatives include the California Leafy Greens Agreement and Global Gap.

The 2004 Bioterrorism Regulation is used for maintenance of records one step forward and one step backward to and from immediate suppliers and distributors. Up to 2 years of records might be required in the event of a food recall.

On 21 December 2010, the United States Congress passed the Food and Drug Administration (FDA) Food Safety Modernization Act, the first major food safety measure in the United States in more than 60 years. This new measure will provide more power for the United States FDA to better regulate produce. For example, the FDA will be able to issue direct recalls, instead of relying on voluntary recalls from individual firms or stakeholders. The measure will require importers to be responsible for making sure the food they bring into the country meets United States safety standards. It will require all food participants to take preventive measures and to undergo more random inspections. The legislation also deals with the issue of mass production and centralization that now dominates the food industry, providing new requirements for food testing and tracking (traceability).

Table 14.1. Increasing United States food imports and trade (US$).

Food group	1999	2000	2001	2002	2003	2004	2005	2006	2007	2008	2009
Total US food imports[a]	41,161.8	43,080.7	43,641.8	46,558.6	52,302.6	58,219.6	64,202.1	70,587.2	76,563.0	84,276.4	76,839.4
Live farm animals	1,245.7	1,468.9	1,833.0	1,787.8	1,290.1	1,115.2	1,675.1	2,175.8	2,597.3	2,297.7	1,653.0
Meats	3,276.7	3,838.2	4,270.9	4,298.4	4,439.6	5,736.6	5,774.7	5,268.9	5,403.3	5,110.1	4,648.5
Fish and shellfish	8,877.6	9,902.5	9,685.8	9,990.8	10,895.0	11,145.2	11,890.1	13,176.2	13,508.7	14,024.0	13,031.9
Dairy	930.3	922.3	995.8	1,008.8	1,110.3	1,292.3	1,388.4	1,405.7	1,500.8	1,595.6	1,353.0
Vegetables	3,589.2	3,727.5	4,124.0	4,345.0	5,030.6	5,689.7	6,006.1	6,576.4	7,209.6	7,748.4	7,449.6
Fruits	4,793.5	4,647.1	4,677.2	5,080.0	5,567.2	5,974.4	6,907.8	7,735.0	9,250.9	9,913.1	9,657.3
Nuts	794.3	809.3	671.3	701.9	780.1	1,082.4	1,124.9	1,105.3	1,194.6	1,363.1	1,285.9
Coffee and tea	3,108.7	2,921.3	1,915.7	1,942.2	2,227.3	2,559.7	3,309.3	3,695.5	4,171.4	4,855.3	4,508.6
Cereals and bakery	2,659.9	2,735.0	2,990.6	3,343.9	3,619.2	4,012.6	4,242.9	4,911.6	5,916.9	7,691.8	6,847.6
Vegetable oils	1,553.6	1,525.0	1,316.7	1,446.3	1,689.9	2,488.0	2,626.1	3,066.7	3,853.4	6,178.9	4,391.4
Sugar and candy	1,593.7	1,560.9	1,606.6	1,860.4	2,137.6	2,114.3	2,500.6	3,055.3	2,606.8	2,990.5	3,100.8
Cocoa and chocolate	1,521.9	1,403.0	1,535.0	1,759.8	2,438.2	2,483.5	2,750.0	2,658.8	2,661.6	3,298.4	3,476.0
Other edible products	2,498.6	2,514.8	2,625.3	2,864.2	4,134.9	5,172.9	5,800.1	6,245.7	6,418.0	7,031.7	6,445.2
Beverages	4,718.1	5,104.9	5,394.0	6,129.3	6,942.7	7,352.9	8,206.1	9,510.4	10,269.8	10,177.8	8,990.6
Liquors	2,381.9	2,725.5	2,847.0	3,091.1	3,438.2	3,709.2	4,089.9	4,511.5	5,047.7	5,040.4	4,788.4
Total animal foods	14,330.3	16,131.9	16,785.4	17,085.7	17,735.0	19,289.2	20,728.3	22,026.5	23,010.1	23,027.4	20,686.4
Total plant foods	22,113.4	21,843.9	21,462.4	23,343.6	27,625.0	31,577.5	35,267.7	39,050.3	43,283.1	51,071.2	47,162.4
Total beverages	7,099.9	7,830.5	8,241.0	9,220.4	10,380.9	11,062.1	12,296.0	14,021.9	15,317.4	15,218.3	13,779.1
Total US agricultural imports	37,672.8	38,974.5	39,366.0	41,915.3	47,383.7	53,989.2	59,291.1	65,325.8	71,913.0	80,487.7	71,698.8
Non-food agricultural imports	5,388.6	5,796.2	5,410.0	5,347.5	5,976.1	6,914.7	6,979.0	7,914.7	8,858.7	10,235.4	7,891.3

Based on USDA data (http://apps.fas.usda.gov/gats/default.aspx).
[a]Values are obtained by calendar year. Total food imports exclude liquors.

Technology at POEs used could include:

- X-ray and gamma-imaging systems. Customs Border Protection (CBP) officers analyze these images to determine anomalies associated with the cargo listed on the manifest.
- Automated targeting system (ATS). This enables CBP officers to collect and analyze cargo shipping data, to distinguish and select high-risk shipments for further review and examination. ATS is a mathematical model that uses weighted rules that assign a risk score to each arriving shipment in a container, based on manifest information.
- The United States Department of Agriculture Animal and Plant Health Inspection Service (USDA-APHIS) initiates inspections of imported produce at the farms in Mexico or their packing and processing facilities with a 24-hour e-manifest rule (United States Department of Homeland Security, 2003).
- The CBP uses intelligence and risk-based strategies to screen information on 100% of the cargo before it is loaded on to vessels bound for the United States.

Layered defense

- Screening and inspection: CBP screens 100% of all cargo before it arrives in the United States using intelligence and cutting-edge technologies. CBP inspects all high-risk cargo.
- 24-Hour rule: Manifest information must be provided 24 hours prior to the sea container being loaded on to the vessel in the foreign port. CBP may deny the loading of high-risk cargo.
- C-TPAT/FAST: CBP created a public–private and international partnership with over 6,000 businesses. CBP and the companies work together to baseline security standards for supply chain and container security. Customs–Trade Partnership Against Terrorism (C-TPAT) is a joint voluntary government–business initiative to build cooperative relationships that strengthen and improve overall international supply chain operations and United States border security. It was implemented after the 11 September terrorist attacks. It was launched in November 2001, with just seven major United States importers. As of April 2005, there were more than 9,000 participating companies (Michael Chertoff, United States Department of Homeland Security).
- The Free and Secure Trade (FAST) Program seeks to expedite the clearance of trans-border shipments by reducing customs information requirements, dedicating lanes at major crossings to FAST participants, using common technology, and physically examining low-risk shipments with minimal frequency. One major benefit of C-TPAT and FAST is to increase information gain and reduced border delay times.
- CARVER plus Shock is a risk-assessment survey instrument (http://www.fda.gov/downloads/Food/FoodDefense/FoodDefensePrograms/UCM376929.pdf) used to assess risks and vulnerabilities to food risks.

Some attributes of the 2010 Food Modernization Act

The 2010 Food Safety Modernization Act is based on the premise that HACCP may not be sufficient, although it is a good start. It includes: specificity with suppliers other stakeholders, foreign facilities, and refusal of inspection (FSMA § 306); foreign supplier verification program (FSMA § 301); voluntary qualified importer program (FSMA § 302); import certification (FSMA § 302); whistleblower protection (protects employees who provide information; testify, assist, or participate in a proceeding violation; or object to activity, policy, practice, or assigned task they reasonably believe to be a violation of safety protocols); supplier agreements; insurance audit; FDA inspection plan; food safety plan; plan for import compliance; and traceability and recall plan (Civic Impluse, 2016).

Comparing three measures to restore trust in global food supply trade

Some attributes of California leafy green standard

The California Leafy Green Products Handler Marketing Agreement (LGMA) was created in 2007 by Californian farmers. The primary objective of LGMA is to protect public health by improving food safety practices throughout the entire supply chain. It includes 14 leafy green products associated with the LGMA program (arugula, baby leaf lettuce, butter lettuce, cabbage, chard, endive, escarole, green leaf lettuce, iceberg lettuce, kale, red leaf lettuce, romaine lettuce, spinach, and spring mix).

Membership is voluntary and open to firms throughout the supply chain including handlers, processors, and distributors. LGMA members represent 99% of the volume of California leafy greens, with approximately 120 members. LGMA-member companies are required to sell and ship only leafy green products that have been produced by farmers that are in compliance with the LGMA-accepted food safety practices. The LGMA Service Mark is carried on all sales documents such as bills of lading and ship manifests, making leafy green products easily recognizable.

The LGMA food safety best practices cover five key areas:

- general requirements;
- environmental assessments;
- water use;
- soil amendments; and
- work practices and field observations.

Some attributes of the United States FDA Guide to Minimize Microbial Food Safety Hazards

The FDA's guide was created in 1998 in response to increased outbreaks of foodborne illness (FDA, 1998). The purpose is to provide the industry with guidelines to minimize microbial food safety hazards for fresh produce during all stages of the farm-to-table food chain, including:

- growers;
- farm workers;
- packers;
- shippers;
- transporters;
- importers;
- wholesalers;
- retailers;
- government agencies; and
- consumers.

This guidance does not establish legally binding requirements; instead, it should be viewed only as a recommendation. Alternative approaches may be used as long as they are consistent with applicable laws and regulations. The guidance document addresses microbial food safety hazards and good agricultural and management practices associated with: growing, harvesting, washing, sorting, packing, and transporting. The guidance addresses microbial food safety hazards and good agricultural and management practices.

The USDA policy involves mitigating the risk of microbial contamination includes five major areas:

- water quality;
- manure/municipal bio-solids;
- worker hygiene;

- field, facility, and transport sanitation; and
- traceability.

Some attributes GLOBALGAP

EurepGAP originated in 1997 by EUREP British retail members and European supermarkets, and was designated GLOBALGAP in 2007. GLOBALGAP acknowledges consumer concerns on food safety, animal welfare, environmental protection, and worker welfare. GLOBALGAP is a single integrated standard with modular applications for crops (fruit and vegetables, flowers and ornamentals, combinable crops, coffee, tea, cotton, etc.), livestock (cattle, sheep, dairy, pigs, poultry, etc.), and aquaculture (salmon, trout, shrimp, tilapia, etc.).

GLOBALGAP is a voluntary global membership, consisting of producer/supplier members who must comply with standards. Once certified, they are able to display the logo and vocalize individual interests in governance structure. Retail and food service members must demonstrate that policies are in line with GLOBALGAP core values. Membership enables the use of the GLOBALGAP database and allows the use of the GLOBALGAP brand in communication strategies. Associate members significantly influence GLOBALGAP standards by setting processes and making contributions to National Technical Working Groups and other consultation processes.

The GLOBALGAP areas of focus include:

- traceability and record keeping;
- record keeping and internal/self-inspection;
- varieties and rootstocks;
- site history management, and soil and substrate management;
- fertilizer use and irrigation;
- crop protection and harvesting;
- produce handling;
- waste/pollution management, recycling and re-use; and
- worker health, safety and welfare, and complaint forms.

Similarities among standards and policies

- *Traceability*: All three programs require a product traceability system.
- *Quality of irrigation water*: The standards and processes ensure irrigation water quality is

adequate for intended use by having records of irrigation water usage, banning the use of untreated sewage water for irrigation, reviewing existing practices, and performing microbial tests.

- *Hygiene*: Processes and personnel are covered by the three programs.
- *Pest management*: It is mandatory that producers adhere to laws and regulations in regards to pest control. GLOBALGAP and USDA include pest management associated with packaging and storage areas. They are required to prevent, monitor, observe, and intervene when pests of significant risk are present.

Overlaps among programs

LGMA

Environmental assessment of the production field and surrounding area is required prior to the first seasonal planting:

- produce field;
- adjacent land use;
- historical land use;
- flooding; and
- harvest practices.

USDA

Awareness is required of current and historical uses of the land.

- Evaluate production area with regard to proximity of adjacent land uses.
- Determine if controls are designed to avoid contamination of agricultural waters from other farm or animal operations.

It does not require environmental assessment. It requires soil management processes:

- prepared soil maps for the farm; and
- use of cultivation techniques to avoid soil compaction and reduce possibility of soil erosion.

LGMA versus USDA

- LGMA and USDA focus on the appropriate storage of organic fertilizers.
- LGMA and USDA acknowledge the use of soil amendments that contain animal manure that are physically heat treated or processed by other equivalent methods:

- Natural fertilizers, such as composted manure.
- Fertilizers containing natural components.
- Non-synthetic crop treatment.

LGMA requires implementation of management plans that ensure that the crop treatments do not pose a significant contamination hazard, while the USDA guidelines do not present recommendations for plant protection.

GLOBALGAP

- Emphasizes the appropriate storage of inorganic fertilizers.
- Does not acknowledge the use of treated organic fertilizer.
- Has the most comprehensive audit process with regard to plant protection products, including:
  - validating choice of plant protection products used;
  - extensive recording keeping of application;
  - requirements of application equipment conditions;
  - disposal of surplus and obsolete application mix;
  - storage conditions and empty containers;
  - handling processes; and
  - residue analysis.

All three programs

- The three guidelines require that field equipment, such as harvesting machinery, knives, containers, tables, baskets, packaging materials, brushes and buckets, are properly used and sanitized before using.
- Growers are expected to ensure produce that is washed, cooled, or packaged in the field is not contaminated in the process.

LGMA

Further considerations of field equipment sanitation include:

- Adequate distance for turning and manipulation of harvest equipment.
- Designing equipment by using materials and construction that facilitate cleaning/sanitation of equipment food contact surfaces.
- Locating equipment cleaning and sanitizing operations away from the product and other equipment to reduce the potential for contamination.
- LGMA does not acknowledge specific standards associated with product storage and packaging areas.
- LGMA does identify sanitary requirements.

GLOBALGAP and USDA

Have more sanitation requirements associated with product packaging and storage areas. These requirements include:

- Handling and storage facilities and equipment are cleaned/maintained to prevent contamination.
- All forklifts and other driven transport trolleys are clean and well maintained.
- Pallets and containers/bins are to be cleaned before using to transport fresh produce.
- Packing materials should be stored in a way that protects them from contamination.

Differences among programs

LGMA

Best business practices include a comprehensive "flooding" section that covers the following topics related to flooding:

- a product that has come into contact with flooded water;
- a product in proximity to flooded area (not in contact with water); and
- a formerly flooded production ground.

USDA

- Identifies systems and practices to ensure safe management and disposal of waste from permanent or portable toilets.
- Provides sanitation standards for containers used to transport or store water for hand washing.
- Provides additional information with regard to post-harvest processes, such as hydrocooling, use of dump tanks, and flume transport.
- Presents different methods to wash different types of produce, including:
  - submersion, spray, or both;
  - vigorous washing of produce not subject to injury;
  - dry cleaning methods used with some produce that cannot tolerate water; and
  - considering the wash water temperature for certain produce.

The use of antimicrobial chemicals for surface treatments and water sanitation section contains information for the following:

- surface treatments with antimicrobial chemicals;
- options for water sanitation most appropriate for individual operations;

- advising that antimicrobial chemical levels should be monitored and recorded to ensure they are maintained at appropriate concentrations.

It includes a transportation section that advises operators and others involved in the transport of fresh produce to examine product transportation throughout the entire supply chain. These steps include:

- transportation from the field to the cooler;
- operations within packing facility; and
- transportation to distribution and wholesale terminal markets or retail centers.

GLOBALGAP

Propagation material is identified during the audit process with regard to the quality of seeds (free from injurious pests, diseases, virus) and includes standards for genetically modified organisms. It is also used to examine post-harvest treatments such as biocides, waxes, and plant protection products used on harvested crop and the application methods. It expands on the pre-harvest treatments section by including:

- application machinery;
- correct mixing, handling, and filling procedures; and
- disposal of surplus application mix and obsolete plant protection products:
  - plant protection product residue analysis;
  - plant protection product storage and handling; and
  - empty plant protection product containers.

Comprehensive evaluation of programs

- LGMA has the most comprehensive best business practices associated with the farm operations (Climate Conditions and Environment, Soil Amendments, Environmental Assessment, Water Usage and Flooding).
- USDA "Guidance for Industry" has the most comprehensive guidelines for the transportation of produce to the final destination.

GLOBALGAP has the most comprehensive audit process with regard to produce packaging and storage.

Summary

With increased international trade on the marketing system and limited resources to inspect and test all products for pathogens, pathogens could spread fairly rapidly across our borders causing significant economic loss. However, public–private partnership with regulation (USDA) and standards (LGMA and GLOBALGAP) create mechanisms to restore trust in food trade across the globe.

References

Buzby, J. and Roberts, T. (1997) Economic costs and trade impacts of microbial foodborne illness. *World Health Statistics Quarterly* 50, 57–66.

Cameron, G., Pate, J., and Vogel, K.M. (2001) Planting fear. How real is the threat of agricultural terrorism? *Bulletin of the Atomic Scientists* 57, 38–44.

Carus, W.S. (1999) *Bioterrorism and Biocrimes: The Illicit Use of Biological Agents in the 20th Century.* Center for Counter-Proliferation Research, National Defense University, Washington, DC.

Chalk, P. (2003) *The Bio-terrorist Threat to Agricultural Livestock and Produce*, CT-213 Testimony, presented before the Government Affairs Committee of the United States Senate, 19 November.

Civic Impulse. (2016). S. 510 – 111th Congress: FDA Food Safety Modernization Act. Available at: https://www.govtrack.us/congress/bills/111/s510 (accessed 19 October 2016).

FDA (1998) *Guide to Minimize Microbial Food Safety Hazards for Fresh Fruits and Vegetables.* Available at: http://www.fda.gov/downloads/Food/Guidance Regulation/UCM169112.pdf (accessed 30 December 2015).

Regmi, A. (2001) *Changing Structure of Global Food Consumption and Trade.* Market and Trade Economics Division, Economic Research Service, United States Department of Agriculture, Agriculture and Trade Report, WRS-01-1.

Schmidt, J. (2007) U.S. food imports outrun FDA resources. *USA Today* March 18.

USDA-FAS (2008) U.S. trade imports – FAS commodity aggregations. FAS United States trade database. Available at: http://www.fas.usda.gov/data (accessed 17 October 2016).

US Department of Homeland Security (2003) Bureau of Customs and Border Protection; 19 CFR Parts 4, 103, *et al.*; Required Advance Electronic Presentation of Cargo Information; Final Rule. Available at: https://www.census.gov/foreign-trade/regulations/fedregnotices/ADVANCEFILING.pdf. (accessed 17 October 2016).

Glossary

The glossary defines terms used in this book. Some definitions come from the United States Department of Agriculture's National Agricultural Library, available at http://agclass.nal.usda.gov/glossary.shtml, and are identified with (USDA NAL). Some definitions come from Oxford Dictionaries, available at www.oxforddictionaries.com, and are identified with (Oxford Dictionaries).

Abortion The expulsion of a fetus from the womb before it is able to survive independently.

Abscess Accumulation of purulent material in tissues, organs, or circumscribed spaces, usually associated with signs of infection.

Accuracy The extent to which a measurement approaches the true value of the measured quantity (USDA NAL).

Acetic acid An organic acid with antimicrobial activity. Acetic acid is the organic acid in vinegar.

Acidified food A low-acid food that has acid added to lower the pH to 4.6 or lower.

Acidity The level of acid in substances such as water, soil, or wine (Oxford Dictionaries).

Acid tolerance response (ATR) The resistance of bacteria to low pH when they have been grown at moderately low pH or when they have been exposed to a low pH for some time.

Acute (Of a disease or its symptoms) severe but of short duration (Oxford Dictionaries).

Adulteration Render (something) poorer in quality by adding another substance (Oxford Dictionaries).

Aerobic Relating to or denoting exercise taken to improve the efficiency of the body's cardiovascular system in absorbing and transporting oxygen (Oxford Dictionaries).

Aflatoxins Any of a class of toxic compounds produced by certain moulds found in food, which can cause liver damage and cancer (Oxford Dictionaries).

Agriculture The science or practice of farming, including cultivation of the soil for the growing of crops and the rearing of animals to provide food, wool, and other products (Oxford Dictionaries).

Agrochemical A chemical used in agriculture, such as a pesticide or a fertilizer (Oxford Dictionaries).

Alarm water Water content that should not be exceeded if mold growth is to be avoided.

Alcohol A colorless volatile flammable liquid that is produced by the natural fermentation of sugars and is the intoxicating constituent of wine, beer, spirits, and other drinks, and is also used as an industrial solvent and as fuel (Oxford Dictionaries).

Algae A simple, non-flowering, and typically aquatic plant of a large assemblage that includes the seaweeds and many single-celled forms. Algae contain chlorophyll but lack true stems, roots, leaves, and vascular tissue (Oxford Dictionaries).

Allergic reaction Altered reactivity to an antigen, which can result in pathologic reactions upon subsequent exposure to that particular antigen (USDA NAL).

Amortize Gradually write off the initial cost of (an asset) over a period (Oxford Dictionaries).

Anisakis simplex A roundworm parasite transmitted to humans by consumption of undercooked infected seafood.

Animalia The kingdom of living organisms that comprises the animals, now usually including only metazoans but formerly also including protozoans (Oxford Dictionaries).

Anaerobic Relating to or requiring an absence of free oxygen (Oxford Dictionaries).

Animal and Plant Health Inspection Service (APHIS) An agency of the United States Department of Agriculture responsible for protecting animal health, animal welfare, and plant health.

Annelid A segmented worm of the phylum Annelida, such as an earthworm or leech (Oxford Dictionaries).

Antibiotics Chemical substances produced by microorganisms or synthetically that inhibit the growth of, or destroy, bacteria. Antibiotics are used at therapeutic levels to fight disease in humans and animals. Since the 1950s, they have been used at subtherapeutic levels in animal feeds to enhance growth and prevent disease in livestock and poultry (USDA NAL).

Antimicrobial Active against microbes (Oxford Dictionaries).

AOAC International An organization focused on uniformity of testing methods.

Appendicitis A serious medical condition in which the appendix becomes inflamed and painful. The appendix is a tube-shaped sac attached to and opening into the lower end of the large intestine in humans and some other mammals (Oxford Dictionaries).

Aquatic (Of a plant or animal) growing or living in or near water (Oxford Dictionaries).

Aspergillus A genus of mold that contains some species and strains that produce mycotoxins, such as aflatoxins, in foods.

Asphyxiation The state or process of being deprived of oxygen, which can result in unconsciousness or death; suffocation (Oxford Dictionaries).

Asymptomatic (Of a condition or a person) producing or showing no symptoms (Oxford Dictionaries).

Bacillus cereus A Gram-positive, spore forming, thermoduric bacteria that can cause intoxication with an emetic toxin preformed in food or a toxin-mediated infection resulting in diarrhea.

Bacteriocins Substances elaborated by specific strains of bacteria that are lethal against other strains of the same or related species. They are protein or lipopolysaccharide–protein complexes used in taxonomy studies of bacteria.

Bacteriological Analytical Manual (BAM) The United States Food and Drug Administration published analytical manual is the preferred laboratory procedures for the detection in food and cosmetic products of pathogens and microbial toxins.

Beaver fever A colloquialism for giardiasis caused by the protozoan parasite *Giardia*.

Bioaccumulate (Of a substance) become concentrated inside the bodies of living things (Oxford Dictionaries).

Binary fission A mechanism of cell division that originates with one mother cell and results in two daughter cells.

Bioinformatics A field of biology concerned with the development of techniques for the collection and manipulation of biological data, and the use of such data to make biological discoveries or predictions. This field encompasses all computational methods and theories applicable to molecular biology and areas of computer-based techniques for solving biological problems including manipulation of models and datasets (USDA NAL).

Biological activity The ability of a chemical to affect physiological functions of a cell or organism. May also refer to a microorganism that is alive and actively metabolizing chemicals.

Biological kingdoms The highest rank of taxonomic category, grouping together all forms of life having certain fundamental characteristics in common. For the purposes of description in this book, the kingdoms are: Animalia (animals), Plantae (plants), Fungi (yeasts, molds, and mushrooms), Protista (protozoa), and Eubacteria. Other definitions exist.

Biosensors Any of a variety of procedures which use biomolecular probes to measure the presence or concentration of biological molecules, biological structures, microorganisms, etc., by translating a biochemical interaction at the probe surface into a quantifiable physical signal (USDA NAL).

Birth defect A physical or biochemical abnormality that is present at birth and that may be inherited or the result of environmental influence (Oxford Dictionaries).

Botanical Having to do with plants or plant biology.

Botulism A potentially life-threatening paralytic intoxication caused by the potent botulinum neurotoxin produced by *Clostridium botulinum*.

Bovine spongiform encephalopathy A transmissible encephalopathy of cattle characterized by a spongiform structure of the brain tissue, and associated with abnormal prion proteins in the brain (USDA NAL).

Bradyzoite An encysted sporozite, such as those of *Toxoplasma gondii*.

Campylobacter A genus of Gram-negative, microaerohilic, and corkscrew-shaped bacteria that contains potentially pathogenic species that cause foodborne infections known as camplylobacteriosis.

Carnivores An organism (plant or animal) that feeds on animal substances (USDA NAL).

Carrier state The condition of harboring an infective organism without manifesting symptoms of infection. The organism must be readily transmissible to another susceptible host (USDA NAL).

Case An instance of a disease, injury, or problem (Oxford Dictionaries).

Causation The relationship between cause and effect; causality: a strong association is not a proof of causation (Oxford Dictionaries).

Cell membrane The semipermeable membrane surrounding the cytoplasm of a cell (Oxford Dictionaries).

Centers for Disease Control and Prevention (CDC) A federal agency that conducts and supports health promotion, prevention, and preparedness activities in the United States with the goal of improving overall public health.

Chill tank A cooling tank in which food items are immersed into cold water to rapidly lower the temperature of the food.

Cholera An infectious and often fatal bacterial disease of the small intestine, typically contracted from infected water supplies and causing severe vomiting and diarrhea (Oxford Dictionaries).

Cholera toxin A protein complex secreted by the bacterium *Vibrio cholerae* that is toxic and can cause massive, watery diarrhea.

Chronic (Of an illness) persisting for a long time or constantly recurring (Oxford Dictionaries).

Class I recall A situation in which there is a reasonable probability that the use of or exposure to a violative product will cause serious adverse health consequences or death.

Class II recall A situation in which use of or exposure to a violative product may cause temporary or medically reversible adverse health consequences or where the probability of serious adverse health consequences is remote.

Class III recall a situation in which use of or exposure to a violative product is not likely to cause adverse health consequences.

Clostridium botulinum A Gram-positive, spore-forming, anaerobic, thermoduric bacteria that can produce a potent neurotoxin in food that causes botulism.

Clostridium perfringens A Gram-positive, spore-forming, anaerobic, thermoduric bacteria that can cause a toxin-mediated infection resulting in diarrhea.

Commensalism An association between two organisms in which one benefits and the other derives neither benefit nor harm (Oxford Dictionaries).

Competitive exclusion The displacement or elimination of a species from its habitat by another species through interspecific competition (USDA NAL).

Competitive microflora Microorganisms that compete with others for niches in microenvironments.

Composts Organic residues, or a mixture of organic residues and soil, which have been piled, moistened, and allowed to undergo biological decomposition for use as a fertilizer (USDA NAL).

Conidia Asexual spores produced by fungi.

Conidiophores (In certain fungi) a conidium-bearing hypha or filament (Oxford Dictionaries).

Consumer A person who purchases goods and services for personal use (Oxford Dictionaries).

Contaminant Material or microorganisms that are not natural to a substance and that can cause harm.

Control measures Measures taken to reduce adverse effects.

Cook (Of food) be heated so that the state required for eating is reached (Oxford Dictionaries).

Cost of illness analysis A determination of the economic impact of a disease or health condition, including treatment costs; this form of study does not address benefits/outcomes (USDA NAL).

Cross-contamination The process by which bacteria or other microorganisms are unintentionally transferred from one substance or object to another, with harmful effect (Oxford Dictionaries).

Crustacean An arthropod of the large, mainly aquatic group Crustacea, such as a crab, lobster, shrimp, or barnacle (Oxford Dictionaries).

Cryptosporidium A parasitic protozoan that causes an infection known as cryptosporidiosis.

Culinary Having to do with cooking or food preparation.

Culturing techniques Microbiological techniques that involve growing microorganisms in controlled environments.

Customs and Border Protection (CPB) The largest federal law enforcement agency of the United States Department of Homeland Security. It is charged with regulating and facilitating international trade, collecting import duties, and enforcing United States regulations, including trade, customs, and immigration.

Cyclospora A protozoan parasite that is transmitted by feces and causes an infection known as cyclosporiasis.

Cysticercosis Infection with cysticercus, the larval form of the various tapeworms of the genus *Taenia* (usually *T. solium* in man). In humans they penetrate the intestinal wall and invade subcutaneous tissue, brain, eye, muscle, heart, liver, lung, and peritoneum. Brain involvement results in neurocysticercosis (USDA NAL).

Cytotoxins Substances that are toxic to cells; they may be involved in immunity or may be contained in venoms. These are distinguished from cytostatic agents in degree of effect. Some of them are used as cytotoxic antibiotics. The mechanism of action of many of these are as alkylating agents or mitosis modulators (USDA NAL).

Dairy products Food and beverages derived from milk.

Death phase The period of a growth curve after the stationary phase where the nutrients are becoming depleted and waste products are building up, and the multiplication rate drops below the death rate.

Decision tree A tree diagram used to represent the various stages of a decision-making process, typically with each node representing a decision or question and each branch representing a possible consequence or answer resulting from the previous node (Oxford Dictionaries).

Definitive host An organism that supports the adult or sexually reproductive form of a parasite.

Dehydration The loss or removal of water from something, also a harmful reduction in the amount of water in the body (Oxford Dictionaries).

Deoxynivalenol (DON) The most abundantly produced mycotoxin within the family of mycotoxins known as trichothecenes. Also known as vomitoxin.

Department of Homeland Security (DHS) A department of the United States federal government charged with protecting United States territory from terrorist attacks and providing a coordinated response to large-scale emergencies.

Depreciation A reduction in the value of an asset over time, due in particular to wear and tear (Oxford Dictionaries).

Detection limit Concentration or quantity that is derived from the smallest measure that can be detected with reasonable certainty for a given analytical procedure (USDA NAL).

Diagnosis The identification of the nature of an illness or other problem by examination of the symptoms (Oxford Dictionaries).

Diarrhea A condition in which faeces are discharged from the bowels frequently and in a liquid form (Oxford Dictionaries).

Digestion The process of digesting food (Oxford Dictionaries).

Digestive tract The series of organs in animals involved with digestion of food. Typically includes the mouth, esophagus, stomach, small intestine, large intestine, rectum, and anus.

DNA fingerprint A technique for identifying individuals of a species that is based on the uniqueness of their DNA sequence (USDA NAL).

Domestic Existing or occurring inside a particular country; not foreign or international (Oxford Dictionaries).

Dose The amount of a chemical or number of microorganisms consumed.

Dose response The relationship between the amount of exposure (dose) to a substance and the resulting changes in physiological function or health (response) (USDA NAL).

Duration The period of time from onset of symptoms until the symptoms end.

Ecological factors The physical, chemical and biological characteristics of an environment that affect organisms in that environment. Also called environmental factors.

Ecology A branch of science concerned with the interrelationships of organisms and their cycles and rhythms, community development and environments – especially as manifested by natural structure, interaction between different kinds of organisms, geographic distributions, and population alterations (USDA NAL).

Economic Resource Service (ERS) An agency within the United States Department of Agriculture with the mission to anticipate trends and emerging issues in agriculture, food, the environment, and rural America and to conduct high-quality, objective economic research to inform and enhance public and private decision making.

Ecosystem A functional system which includes the organisms of a natural community together with their environment (USDA NAL).

Eggs Reproductive excretions of many species within the animal kingdom. Many forms of eggs are commonly consumed by humans including poultry eggs, reptile eggs and roe from fish.

Elimination The expulsion of waste matter from the body (Oxford Dictionaries).

Emetic toxin A substance that causes vomiting.

Encysted larvae Immature forms of certain parasitic animals that are enclosed in a cyst or capsule.

Endemic (Of a disease or condition) regularly found among particular people or in a certain area (Oxford Dictionaries).

Enrichment culture A technique for isolating organisms in which nutritional and/or environmental conditions are controlled to favor the growth of a specific organism or group of organisms (USDA NAL).

Enteric fever Another term for typhoid or paratyphoid (Oxford Dictionaries).

Enterotoxins Substances that are toxic to the intestinal tract causing vomiting, diarrhea, etc.; most common enterotoxins are produced by bacteria (USDA NAL).

Enterovirus Any of a group of RNA viruses (including those causing polio and hepatitis A) which typically occur in the gastrointestinal tract, sometimes spreading to the central nervous system or other parts of the body (Oxford Dictionaries).

Entomophagy The consumption of insects.

Environmental factors The physical, chemical, and biological characteristics of an environment that affect organisms in that environment. Also called ecological factors.

Environmental microorganism Microorganisms that are commonly found in many different types of environments.

Enzyme A substance produced by a living organism which acts as a catalyst to bring about a specific biochemical reaction (Oxford Dictionaries).

Environmental pressures The physical and chemical aspects of a system that influence microbial ecosystems and biological processes.

Epidemiology The study of the various factors influencing the occurrence, distribution, prevention, and control of disease, injury, and other health-related events in a defined population (USDA NAL).

Epithelial cells Cells forming the cellular sheets that cover surfaces, both inside and outside the body.

Escherichia coli A species of Gram-negative, short rod-shaped, bacteria that are commonly found in intestinal tracts of animals. Some subtypes may be pathogenic (see virotypes).

Etiology The branch of science concerned with the causes and origins of diseases (USDA NAL).

Eubacteria A bacterium of a large group typically having simple cells with rigid cell walls and often flagella for movement. The group comprises the "true" bacteria and cyanobacteria, as distinct from archae (Oxford Dictionaries).

Eukaryotes Cells of the higher organisms, containing a true nucleus bounded by a nuclear membrane.

European Mycotoxins Awareness Network (EMAN) A non-profit consortium that exists to provide high-quality scientific information on mycotoxins, such as aflatoxins, ochratoxin A, trichothecenes (including deoxynivalenol or vomitoxin), fumonisins, zearalenone and ergot alkaloids, to industry, consumers, legislators, and the scientific community.

Exponential phase The period of a growth curve where multiplication is most rapid, occurring at a logarithmic rate. Also called the log phase.

Exposure analysis The determination or estimation (qualitative or quantitative) of the magnitude, frequency, duration and route of exposure of a population (USDA NAL).

Extrinsic factors Environmental factors external to a food.

Facultative aerobes A microorganism that can live with or without oxygen present. Same as facultative anaerobes.

Facultative anaerobes A microorganism that can live with or without oxygen present. Same as facultative aerobes.

Fallen fruit Fruit that has fallen to the ground.

False alarm A warning given about something that fails to happen (Oxford Dictionaries).

Fecal contamination Contaminating material originating from feces.

Fecal–oral route Route of transmission of infectious microorganisms that originate in fecal material and transferred to the next host through oral consumption.

Federal register A publication of the United States federal government presenting proposed and final regulations of federal agencies.

Feeds Any non-injurious edible material having nutrient value; may be harvested forage, range or artificial pasture forage, grain, or other processed food for livestock or game animals (USDA NAL).

Feline Relating to or affecting cats or other members of the cat family (Oxford Dictionaries).

Fermented foods Foods that result from fermentation processes, usually through the activity of desirable microorganisms.

Fertilizers Any organic or inorganic material of natural or synthetic origin which is added to soil to provide nutrients, including nitrogen, phosphorus, and potassium, necessary to sustain plant growth (USDA NAL).

Fetus An unborn or unhatched offspring of a mammal, in particular, an unborn human more than eight weeks after conception (Oxford Dictionaries).

Field fungi Fungi that invade food crops in the field. These fungi can damage grain as the seeds develop.

Firm A business organization that sells goods or services to make a profit.

Fixed costs Business costs, such as rent, that are constant whatever the amount of goods produced (Oxford Dictionaries).

Flagellum A slender thread-like structure, especially a microscopic whip-like appendage which enables many protozoa, bacteria, spermatozoa, etc. to swim (Oxford Dictionaries).

Flu-like symptoms Symptoms similar to those of the flu, including fever, shivering, chills, malaise, and body aches.

Fomites Inanimate objects that carry pathogenic microorganisms and thus can serve as the source of infection. Microorganisms typically survive on fomites for minutes or hours. Common fomites include clothing, tissue paper, hairbrushes, and cooking and eating utensils (USDA NAL).

Food Supplies for living organisms to use for energy and structural maintenance. Any living organism can be a potential food source to other living organisms, humans included.

Food adulteration Mixing other matter of an inferior and sometimes harmful quality with food or drink intended to be sold.

Food counterfeiting Deliberate and intentional substitution, addition, tampering, or misrepresentation of food, food ingredients, or food packaging; or false or misleading statements made about a product, for economic gain. Also called food fraud.

Food and Drug Administration (FDA) The United States Food and Drug Administration is a federal agency of the United States Department of Health and Human Services, one of the United States federal executive departments.

Food and feed insecurity A lack of supply and/or access to food and feed.

Foodborne illness Foodborne disease caused by consuming contaminated foods or beverages (USDA NAL).

Food defense The overall process of protecting the food supply from intentional contamination, including preventive measures, surveillance, incident reporting and control (USDA NAL).

Food deserts Food deserts are defined as urban neighborhoods and rural towns without ready access to fresh, healthy, and affordable food. Instead of supermarkets and grocery stores, these communities may have no food access or are served only by fast food restaurants and convenience stores that offer few healthy, affordable food options (USDA NAL).

Food fraud Deliberate and intentional substitution, addition, tampering, or misrepresentation of food, food ingredients, or food packaging; or false or misleading statements made about a product, for economic gain. Also called food counterfeiting.

Food matrix The nutrient and non-nutrient components of foods and their molecular relationships, i.e. chemical bonds, to each other (USDA NAL).

FoodNet The United States foodborne diseases active surveillance network; a branch of the Centers for Disease Control that monitors foodborne disease outbreaks.

Food networks The complex routes by which food moves and is processed and transported.

Food processing plant A processing plant where food ingredients are transformed into new products on an industrial scale.

Food recall Action by a manufacturer or distributor to protect the public from products that may cause harm. A food recall is intended to remove food products from commerce when there is reason to believe the products may be adulterated or misbranded.

Food safety The fitness of a food for human consumption (USDA NAL).

Food Safety Inspection Service (FSIS) An agency under the United States Department of Agriculture responsible for ensuring the truthfulness and accuracy in labeling of meat and poultry products.

Food Safety Modernization Act (FSMA) Signed into law by President Obama on 4 January 2011. It aims to ensure the United States' food supply is safe by shifting the focus of federal regulators from responding to contamination to preventing it. The United States Food and Drug Administration (FDA) has oversight of this law.

Food scare An instance of widespread public anxiety about the food supply, especially concerning contamination or shortages (Oxford Dictionaries).

Food security Access by all people, at all times to sufficient food for an active and healthy life. Food security includes at a minimum: the ready availability of nutritionally adequate and safe foods, and an assured ability to acquire acceptable foods in socially acceptable ways (USDA NAL).

Food service The practice or business of making, transporting, and serving or dispensing prepared foods, as in a restaurant or school (USDA NAL).

Food spoilage Varying degrees of physical, chemical and/or biological deterioration of food sensory properties, nutrient content, and/or safety. Major causes include: 1) growth and activity of microorganisms, 2) natural enzymes, 3) insects, parasites and rodents, 4) temperature, 5) moisture and dryness, 6) air/oxygen, 7) light, 8) time (USDA NAL).

Food supply chain The processes by which food moves from production to consumer, or in other words from farm to fork.

Food supply networks Similar to food supply chain or food network. The word network implies non-linear complexity.

Food systems All processes and infrastructure involved in feeding a population: growing, harvesting, processing, packaging, transporting, marketing, consumption, and disposal of food and food-related items.

Freeze (Of food) be able to be preserved at a very low temperature (Oxford Dictionaries).

Fresh produce Raw fruits and vegetables (USDA NAL).

Fruit In culinary terms, the sweet and fleshy part of a plant. In botanical terms, a collective plant structure containing one or more embryos, which as a whole, develops from a gynoecium (USDA NAL).

Fumonisins A group of mycotoxins produced by species of *Fusarium* molds.

Fungi The plural form for fungus. A fungus is any of a group of unicellular, multicellular, or syncytial spore-producing organisms feeding on organic matter, including molds, yeast, mushrooms, and toadstools (Oxford Dictionaries).

Fusarium A genus of mold that contains species and strains that are pathogenic to plant crops and may produce mycotoxins such as trichotecenes and fumonisins.

Fusarium **head blight** A destructive plant disease caused by mold species of *Fusarium*. Also known as scab.

Gastroenteritis Inflammation of the stomach and intestines, typically resulting from bacterial toxins or viral infection and causing vomiting and diarrhea (Oxford Dictionaries).

Genetic diversity Refers to the diversity of genetic makeup within a species. A way to describe genetic differences within a species.

Germinate To begin to grow, as in spore germination where the dormant spore comes out of dormancy, becomes biologically active and starts to multiply.

Giardia A flagellate protozoan parasite that causes an infection known a giardiasis.

Good agricultural practices Guidelines and methods for farmers, growers, food producers, and those involved in agriculture to manage the resources important to agriculture, such as soil, water, and air, to produce safe and hygienic food products with commitment to sustainability and conservation, while protecting human and animal health and welfare as well as the environment (USDA NAL).

Good manufacturing practices (GMP) A system of ensuring production consistency with minimal risks.

Grain Wheat or any other cultivated cereal crop used as food (Oxford Dictionaries).

Grain Inspection, Packers and Stockyards Program (GIPSA) An agency of the United States Department of Agriculture that facilitates the marketing of livestock, poultry, meat, cereals, oilseeds, and related agricultural products, and promotes fair and competitive trading practices for the overall benefit of consumers and United States agriculture.

Gram stain A staining technique for the preliminary identification of bacteria, in which a violet dye is applied, followed by a decolorizing agent and then a red dye. The cell walls of certain bacteria (denoted Gram-positive) retain the first dye and appear violet, while those that lose it (denoted Gram-negative) appear red (Oxford Dictionaries).

Growth curve A graphical representation of growth patterns. A bacterial growth curve has phases of growth: lag, log, stationary, and death.

Guillain–Barré syndrome An acute inflammatory autoimmune neuritis caused by T cell-mediated cellular immune response directed towards peripheral myelin. Demyelination occurs in peripheral nerves and nerve roots. The process is often preceded by a viral or bacterial infection, surgery, immunization, lymphoma, or exposure to toxins. Common clinical manifestations include progressive weakness, loss of sensation, and loss of deep tendon reflexes. Weakness of respiratory muscles and autonomic dysfunction may occur (USDA NAL).

Gundi A small gregarious rodent living on rocky outcrops in the deserts of North and East Africa (Oxford Dictionaries).

HACCP Hazard analysis critical control point. A systematic approach to be used in food processing as a means to assure food safety (USDA NAL).

Halophile An organism, especially a microorganism that grows in or can tolerate saline conditions (Oxford Dictionaries). Salt loving.

Hazard A potential source of danger (Oxford Dictionaries).

Hazard assessment A description of what might go wrong and how it might happen.

Hazard identification The description of the types and nature of adverse effects caused by substances, activities or events (USDA NAL).

Heat labile A substance, often a protein, that loses biological activity once heated.

Heat stable A substance that keeps its biological activity after being heated.

Heavy metals Metals with high specific gravity, typically larger than 5. They have complex spectra, form colored salts and double salts, have a low electrode potential, are mainly amphoteric, yield weak bases and weak acids, and are oxidizing or reducing agents (USDA NAL).

Hepatitis A Inflammation of the liver in humans caused by a member of the hepatovirus genus, human hepatitis A virus. It can be transmitted through fecal contamination of food or water (Oxford Dictionaries).

Home canning The process of preserving foods by packing them into glass jars and then heating and hermetically sealing the jars.

Homeostasis The tendency towards a relatively stable equilibrium between interdependent elements, especially as maintained by physiological processes (Oxford Dictionaries).

Homogeneity The quality or state of being all the same or all of the same kind (Oxford Dictionaries).

Host cells Cells from a host.

Host adapted Certain pathogens can only infect specific types of hosts or host cells.

Host susceptibility The likelihood of a particular host becoming infected with a pathogen.

Host utilization Where a microorganism gains sustenance from a host organism.

Hot spot Pockets of biological activity, usually in stored grain, produced by moisture pockets, insects, mites, fungi, and other microorganisms, where the heat produced can actually char the material.

Human host Where a human may host other living microorganisms.

Hygiene Conditions or practices conducive to maintaining health and preventing disease, especially through cleanliness (Oxford Dictionaries).

Immune cell Any cell involved in an immune response or forming part of the immune system (Oxford Dictionaries).

Immune system The organs and processes of the body that provide resistance to infection and toxins. Organs include the thymus, bone marrow, and lymph nodes (Oxford Dictionaries).

Immunocompromised population A population of individuals with a weakened immune system, making them susceptible to illness or infection (USDA NAL).

Immunosensors Analytical devices that use antibodies as the specific sensing element and detect concentration dependent signals (USDA NAL).

Import Bring (goods or services) into a country from abroad for sale (Oxford Dictionaries).

Incubation period The time from the moment of inoculation (exposure to the infecting organism) to the appearance of clinical manifestations of a particular infectious disease (USDA NAL).

Infant botulism A form of botulism in infants that is caused by germination of *Clostridium botulinum* spores that colonize in the intestinal tract and produce botulinum toxin.

Infection State of illness caused by a microorganism infecting a host.

Infective dose The number of pathogenic microorganisms necessary to cause infection in a host.

Inflammation A pathological process characterized by injury or destruction of tissues caused by a variety of cytologic and chemical reactions. It is usually manifested by typical signs of pain, heat, redness, swelling, and loss of function (USDA NAL).

Ingestion The act of taking food, beverages or other substances into the body by mouth (USDA NAL).

Inoculate To introduce microorganisms to a new environment such as a culture medium or food.

Input–output (I–O) model A quantitative economic technique that represents the interdependencies between different branches of a national economy or different regional economies.

Intermediate host An organism that supports the immature or non-reproductive forms of a parasite (Oxford Dictionaries).

Intrinsic factors Environmental factors inherent or internal to a food.

Insect A small arthropod animal that has six legs and generally one or two pairs of wings (Oxford Dictionaries).

Insect infestation Being invaded or overrun by insects.

Inspection Careful examination or scrutiny (Oxford Dictionaries)

International Commission on Microbiological Specifications for Foods (ICMSF) A group of experts initially formed in 1962 to provide timely, science-based guidance to government and industry on appraising and controlling the microbiological safety of foods.

Intoxication Illness caused by consumption of a toxic chemical.

Irradiate Expose (food) to gamma rays to kill microorganisms (Oxford Dictionaries).

Irrigation water Water used for the purpose of irrigating plant crops. Microbial contamination of irrigation water is a concern for food safety.

Jaundice A clinical manifestation of hyperbilirubinemia, characterized by yellowish staining of the skin; mucous membrane; and sclera. Clinical jaundice usually is a sign of liver dysfunction (Oxford Dictionaries).

Kinetics The study of rates of chemical reactions or microbial growth.

Lactic acid An organic acid that has preservative effects due to antimicrobial activity.

Lactic acid bacteria A categorization of Gram-positive, non-spore-forming, beneficial bacteria that produce lactic acid through fermentation.

Lactic acid fermentation Fermentation carried out by lactic acid bacteria in which sugar is converted either entirely, or almost entirely, to lactic acid or to a mixture of lactic acid and other products (USDA NAL).

Lag phase The initial period of a growth curve where bacteria are adjusting to a new environment and are not actively multiplying.

Legume A leguminous plant (member of the pea family), especially one grown as a crop (Oxford Dictionaries).

Leukoencephalomalacia A disease state with symptoms of ataxia, tremor, circling, depressed consciousness, recumbency, and death in horses and donkeys; caused by fumonisins, usually ingested with moldy corn.

LGMA The California Leafy Green Marketing Agreement. LGMA members are companies that ship and sell California-grown lettuce, spinach, and other leafy greens products.

Life cycle The series of changes in the life of an organism including reproduction (Oxford Dictionaries).

Life sciences The sciences concerned with the study of living organisms, including biology, botany, zoology, microbiology, physiology, biochemistry, and related subjects. Often contrasted with physical sciences (Oxford Dictionaries).

Listeria monocytogenes A Gram-positive, facultatively anaerobic, psychrotrophic, potentially pathogenic bacterium that can cause the infection listeriosis.

Livestock Animals that are used and/or produced for agricultural purposes.

Log phase The period of a growth curve where multiplication is most rapid, occurring at a logarithmic rate. Also called the exponential phase.

Loosely coupled complex supply system Loose coupling is an approach to interconnecting the components in a system or network so that those components, also called elements, depend on each other to the least extent practicable. Coupling refers to the degree of direct knowledge that one element has of another. Loosely coupled systems are less likely to be traceable because information is more likely to be ambiguous and less likely to be visible to and understood by the various actors in the system. A complex system can be hypersensitive to small changes. If the rules of interaction and behavior are linear, then the aggregate behavior of the system is linear and not complex.

Loosely coupled linear supply system Loose coupling is an approach to interconnecting the components in a system or network so that those components, also called elements, depend on each other to the least extent practicable. Loosely coupled systems are less likely to be traceable because information is more likely to be ambiguous and less likely to be visible to and understood by the various actors in the system. Coupling refers to the degree of direct knowledge that one element has of another. A complex system can be hypersensitive to small changes.

Low-acid foods Food with pH values higher than 4.6 (USDA NAL).

Lymph nodes Each of a number of small swellings in the lymphatic system where lymph is filtered and lymphocytes are formed (Oxford Dictionaries).

Lyophilization Creation of a stable preparation of a biological substance or culture material by rapid freezing and dehydration of the frozen product under vacuum.

Malaise A general feeling of discomfort, illness, or unease whose exact cause is difficult to identify (Oxford Dictionaries).

Manufacture Make (something) on a large scale using machinery (Oxford Dictionaries).

Marine mammal Aquatic mammals that rely on the ocean and other marine ecosystems for their existence. They include animals such as seals, whales, manatees, sea otters, and polar bears.

Mastitis An infection and inflammation of the udder in cows (USDA NAL).

Mathematical modeling Using mathematical structures such as graphs and equations to represent real-world situations.

Mechanical damage Physical damage to tissue, breaking down natural barriers to microbial transmission.

Melamine A white crystalline compound made by heating cyaniamid and used in making plastics. This chemical has been a food adulterant used to artificially raise the results for protein analysis of food ingredients, and has proven toxic to humans and animals.

Meningitis A serious disease in which there is inflammation of the meninges, caused by viral or bacterial infection, and marked by intense headache and fever, sensitivity to light, and muscular rigidity (Oxford Dictionaries).

Mesophile Microorganisms that are warm loving and can grow in the temperature range of 20–45°C, with an optimum temperature range of 30–40°C.

Methicillin-resistant *Staphylococcus aureus* (MRSA) Strains of *Staphylococcus aureus* that are resistant to the antibiotic methicillin.

Microaerophilic (Of a microorganism) requiring little free oxygen, or oxygen at a lower partial pressure than that of atmospheric oxygen (Oxford Dictionaries).

Microbial-based foods Foods that are derived from microorganisms.

Microbial food safety Practices to manage risk of foodborne illness.

Microbiome The full collection of microbes (bacteria, fungi, virus, etc.) that naturally exist within a particular biological niche such as an organism, soil, a body of water, etc. (USDA NAL)

Microorganism A microscopic organism, especially a bacterium, virus, or fungus (Oxford Dictionaries).

Microscopic–macroscopic (MICMAC) A specific scenario computer program.

MicroVal A European certification organization for the validation and approval of alternative methods for the microbiological analysis of food and beverages.

Modified atmosphere packaging A packaging technique where the gas composition surrounding the product is changed in order to prolong shelf life and reduce natural deterioration, such as gas permeable packaging materials or gas flushing with nitrogen (USDA NAL).

Moisture content The amount of water in a food.

Mold A fungus with visible downy mycelium upon which powdery conidia can be seen (USDA NAL).

Molecular techniques Testing methods that are biochemically based and do not necessarily involve the growth of the target microorganism.

Mollusks Aquatic invertebrate organisms such as oysters and clams in the phylum Mollusca that typically have a body enclosed in a firm, calcareous shell (USDA NAL).

Monocyte A large phagocytic white blood cell with a simple oval nucleus and clear, greyish cytoplasm (Oxford Dictionaries).

Mononucleosis An abnormally high proportion of monocytes in the blood, especially associated with glandular fever (Oxford Dictionaries).

Mortality rate The number of deaths in a given area or period, or from a particular cause (Oxford Dictionaries).

Multi-ingredient processed food A complex mixture of multiple ingredients as a result of processing several ingredients together to produce a food product.

Municipal water A usually treated water supply provided from a central point, piped to individual users.

Mushroom A fungal growth that typically takes the form of a domed cap on a stalk, often with gills on the underside of the cap (Oxford Dictionaries).

Mutualism Symbiosis that is beneficial to both organisms involved.

Myalgia Pain in a muscle or group of muscles (Oxford Dictionaries).

Mycotoxicosis A condition of toxicosis produced by a fungus; poisoning caused by a fungus (Oxford Dictionaries).

Mycotoxins Toxic compounds produced by fungi (USDA NAL).

National Agriculture Release Program (NARP) On 8 January 2007, United States Customs and Border Protection implemented the National Agriculture Release Program. NARP provides a methodology for evaluating high-volume agriculture imports that are low-risk for the introduction of plant pests and plant diseases into the United States.

Neural receptors Receptors embedded in the cell membranes of some nervous system cells that bind to signal chemicals and initiate electrical signals that result in a nervous system response.

Neurodegenerative diseases Hereditary and sporadic conditions which are characterized by progressive nervous system dysfunction. These disorders are often associated with atrophy of the affected central or peripheral nervous system structures (Oxford Dictionaries).

Neurotoxin A toxin that affects the nervous system.

Norovirus Any of various single-stranded RNA viruses including the Norwalk virus and closely related viruses (Oxford Dictionaries).

North American Free Trade Agreement (NAFTA) An agreement among the United States, Canada, and Mexico designed to remove tariff barriers between the three countries.

Nuts A fruit consisting of a hard or tough shell around an edible kernel (Oxford Dictionaries).

Ochratoxin Any of a series of mycotoxins which are produced by various moulds of the genera *Aspergillus* and *Penicillium* and are derivatives of 1-phenylalanine and a compound related to coumarin; especially ochratoxin A, produced chiefly by *A. ochraceus* and having the empirical formula $C_{20}H_{18}ClNO_6$ (Oxford Dictionaries).

Oilseed Any of several seeds from cultivated crops yielding oil, e.g., rape, peanut, soybean, or cotton (Oxford Dictionaries).

Oligopoly A state of limited competition, in which a market is shared by a small number of producers or sellers (Oxford Dictionaries).

Omnivores An organism that feeds on both animal and plant substances (USDA NAL).

Onset time The time between consumption of a pathogen or toxin and the appearance of symptoms of illness.

Oocysts Zygote-containing cysts of sporozoan protozoa. Further development in an oocyst produces small individual infective organisms called sporozoites. Then, depending on the genus, the entire oocyst is called a sporocyst or the oocyst contains multiple sporocysts encapsulating the sporozoites (USDA NAL).

Operating costs Expenses of operating a business. Generally considered expenses on equipment or material that have a life span equal to or less than 1 year.

Opportunistic pathogen An organism that can only cause an infection once the host's resistance is lowered.

Packaging Materials used to wrap or protect goods (Oxford Dictionaries).

Paralysis The loss of the ability to move (and sometimes to feel anything) in part or most of the body, typically as a result of illness, poison, or injury (Oxford Dictionaries).

Parasite An organism that lives in or on another organism (its host) and benefits by deriving nutrients at the host's expense (Oxford Dictionaries).

Pasteurization Process of heating milk or other liquids to destroy microorganisms that can cause disease or spoilage while minimizing chemical changes that affect taste or aroma (USDA NAL).

Pathogen A bacterium, virus, or other microorganism that can cause disease (Oxford Dictionaries).

Pathogen environmental monitoring (PEM) An ongoing measure of the effectiveness of the overall pathogen control program in a food processing plant.

Pathogenicity Ability of infectious agents (i.e., viruses, bacteria, fungi), parasitic nematodes, protozoa or helminths to cause disease (USDA NAL).

Pathotype Populations of the same species which differ by their pathogenic capability (USDA NAL). Same as virotypes or pathovars.

Pathovars Populations of the same species which differ by their pathogenic capability (USDA NAL). Same as pathotypes or virotypes.

Patulin A heterocyclic antibiotic compound, $C_7H_6O_4$, which has carcinogenic properties and is obtained from any of several moulds (originally from *Penicillium patulum*) (Oxford Dictionaries). A type of mycotoxin.

Penicillium A blue mould that is common on food, being added to some cheeses and used sometimes to produce penicillin (Oxford Dictionaries). Some species and strains may produce mycotoxins such as ochratoxin and patulin.

Peptidoglycan A substance forming the cell walls of many bacteria, consisting of glycosaminoglycan chains interlinked with short peptides (Oxford Dictionaries).

Pest control The elimination or control of an insect or animal pest (Oxford Dictionaries). Microorganisms and invasive plants can also be pests that require control.

Pesticides A pesticide is any substance or mixture of substances intended for preventing, destroying, repelling, or mitigating any pest (USDA NAL).

Pests A general term for organisms which may cause illness or damage to humans or to crops, livestock, or materials important to humans (USDA NAL).

pH A measure of the hydrogen-ion activity in solution, expressed on a scale of 0 (highly acid) to 14 (highly basic); pH 7.0 is a neutral solution, neither acid nor basic.

Phagosomes Membrane-bound cytoplasmic vesicles formed by invagination of phagocytized material. They fuse with lysosomes to form phagolysosomes in which the hydrolytic enzymes of the lysosome digest the phagocytized material (USDA NAL).

Pharyngitis Inflammation of the pharynx usually caused by allergens, irritants or infections (USDA NAL).

Phialide A flask-shaped specialized end of a conidiophore from which conidia are formed.

Phyllosphere The three-dimensional microenvironment surrounding a leaf (USDA NAL).

Placenta The placenta consists of vascular tissue in which oxygen and nutrients can pass from the mother's blood into that of the fetus, and waste products can pass in the reverse direction (Oxford Dictionaries).

Plague An acute infectious disease caused by *Yersinia pestis* that affects humans, wild rodents, and their ectoparasites. This condition persists due to its firm entrenchment in sylvatic rodent–flea ecosystems throughout the world. Bubonic plague is the most common form (USDA NAL).

Plantae The biological kingdom of the plants.

Plant pathogen Microorganism that can cause a diseased state in a plant.

Population growth An increase in numbers of a specific group of organisms.

Port of entry (POE) A harbour or airport where customs officers are stationed to oversee people and goods entering or leaving a country (Oxford Dictionaries).

Post-harvest The handling of food crops or animals immediately after the harvest step.

Post-processing contamination Contamination that occurs after a food has been processed.

Potentially hazardous food (PHF) A natural or synthetic food that requires temperature control because it is in a form capable of supporting the rapid and progressive growth of infectious or toxigenic microorganisms (USDA NAL).

Precision The agreement between the numerical values of two or more measurements that have been made in an identical fashion (USDA NAL).

Predictive modeling Use of mathematical models to create predictions of future behaviors.

Pre-harvest Occurring before a crop or livestock for food are harvested.

Premature birth A birth that takes place more than 3 weeks before the due date.

Prevention The action of stopping something from happening or arising (Oxford Dictionaries).

Primates A mammal of an order that includes the lemurs, bushbabies, tarsiers, marmosets, monkeys, apes, and humans. They are distinguished by having hands, hand-like feet, and forward-facing eyes, and are typically agile tree-dwellers (Oxford Dictionaries).

Prions Small proteinaceous infectious particles which resist inactivation by procedures that modify nucleic acids and contain an abnormal isoform of a cellular protein which is a major and necessary component. The abnormal (scrapie) isoform is PrPSc; the cellular isoform is PrPC. The primary amino acid sequence of the two isoforms is identical. Among diseases caused by prions are scrapie, bovine spongiform encephalopathy, and Creuzfeldt–Jakob Syndrome (USDA NAL).

Probiotics Live, nonpathogenic, nontoxic microbial organisms which, when administered in adequate amounts, confer a health benefit on the host (USDA NAL).

Processed food Foods that are manufactured, usually on a large scale, using any of a wide variety of processing techniques, most often with the goal of preserving food for market (USDA NAL).

Processing Perform a series of mechanical or chemical operations on (something) in order to change or preserve it (Oxford Dictionaries).

Producer A person, company, or country that makes, grows, or supplies goods or commodities for sale (Oxford Dictionaries).

Prokaryotes Cells, such as those of bacteria and blue-green algae, which lack a nuclear membrane so that the nuclear material is either scattered in the cytoplasm or collected in a nucleoid region.

Protein Any of a class of nitrogenous organic compounds that consist of large molecules composed of one or more long chains of amino acids and are an essential part of all living organisms, especially as structural components of body tissues such as muscle, hair, collagen, etc., and as enzymes and antibodies (Oxford Dictionaries).

Protista The biological kingdom of the protozoa.

Protozoa A phylum or grouping of phyla which comprises the single-celled microscopic animals, which include amoebas, flagellates, ciliates, sporozoans, and many other forms. They are now usually treated as a number of phyla belonging to the kingdom Protista (Oxford Dictionaries).

Pruritus Severe itching of the skin, as a symptom of various ailments (Oxford Dictionaries).

Psychrophile Microorganisms that are cold loving and can grow below –20°C, with an optimum temperature range of –20 to 0°C. These are environmental extremophiles and are not typically associated with foods.

Psychrotroph Microorganisms that can grow at temperatures below 7°C with an optimum temperature range of 20–30°C. This means they can grow at refrigeration temperatures, but will grow slowly (longer lag phase at lower temperatures).

Public health Branch of medicine concerned with the prevention and control of disease and disability, and the promotion of physical and mental health of the population on the international, national, state, or municipal level (USDA NAL).

Public health emergency An occurrence or imminent threat of an illness that threatens members of the public.

Pulmonary edema Fluid accumulation in the lungs.

PulseNet A network run by the Centers for Disease Control and Prevention (CDC), which brings together public health and food regulatory agency laboratories around the United States.

Quality loss function A mathematical formula developed by Genichi Taguchi in Japan. The result is listed in monetary terms. A formula that estimates the loss of quality that occurs as the result of a product varying from the desired quality.

Quorn A type of protein-rich food made from an edible fungus and used as a meat substitute (Oxford Dictionaries).

Radio frequency identification (RFID) The use of electromagnetic or electrostatic coupling in the radio frequency portion of the electromagnetic spectrum to identify an object, animal, or person.

Rapidity The speed by which a testing method can be completed.

Rapid methods Testing methods that can be completed faster than traditional culturing methods.

Rash An area of redness and spots on a person's skin, appearing especially as a result of illness (Oxford Dictionaries).

Reactive arthritis Joint pain and swelling triggered by an infection.

Ready-to-eat food Food that is in a form ready for consumption and assumed to have a low level of food safety risk.

Recall (Of a manufacturer) request all the purchasers of (a certain product) to return it, as the result of the discovery of a fault (Oxford Dictionaries).

Redox potential (E_h) Measure of the tendency of a chemical species to acquire electrons and thereby be reduced. Reduction potential is measured in volts (V) or millivolts (mV).

Relative humidity Ratio of the amount of water vapor present in the air to that which the air would hold at saturation at the same temperature. It is usually considered on the basis of the weight of the vapor but, for accuracy, should be considered on the basis of vapor pressures (USDA NAL).

Reptile A vertebrate animal of a class that includes snakes, lizards, crocodiles, turtles, and tortoises. They are distinguished by having a dry scaly skin and typically laying soft-shelled eggs on land (Oxford Dictionaries).

Retail The sale of goods to the public in relatively small quantities for use or consumption rather than for resale (Oxford Dictionaries).

Risk The probability that an adverse event will occur, including measures of the probability of an unfavorable outcome (USDA NAL).

Risk analysis The analytical process for assessing, managing and communicating information about potential hazards or other undesirable events, and quantifying their probability and expected consequences (USDA NAL).

Risk assessment The qualitative or quantitative evaluation of the likelihood of adverse effects. A risk assessment generally has four steps: hazard identification, hazard characterization, exposure assessment and risk characterization (USDA NAL).

Risk communication An educational process where potential risks are communicated to individuals or groups of individuals in order to promote risk awareness and prevent adverse incidents (USDA NAL).

Risk management The process an organization undertakes to minimize financial, health or other risks by identifying potential hazards or adverse effects and by planning for and handling incidents which do occur in such a manner that their effect and cost are minimized (USDA NAL).

Risk perception The subjective judgement that people make about the characteristics and severity of a risk.

RNA virus A virus in which the genetic information is stored in the form of RNA (as opposed to DNA) (Oxford Dictionaries).

Roe Fish eggs.

Rotavirus Any of a group of RNA viruses, some of which cause acute enteritis in humans (Oxford Dictionaries).

Ruggedness The ability of a testing method to remain functional under varying conditions.

Salmonella Gram-negative, small rod-shaped bacteria that can cause infections known as salmonellosis.

Sampling plan Outline of which measurements will be taken at what times, on which material, in what manner, and by whom.

Sanitary and phytosanitary regulations Government standards to protect humans, animals and plants from diseases, pests or contaminants (USDA NAL).

Sanitation The development and application of measures designed to promote health and prevent disease (USDA NAL).

Saprophyte A plant, fungus, or microorganism that lives on dead or decaying organic matter (Oxford Dictionaries).

Scab Wheat scab is a destructive plant disease caused by mold species of *Fusarium*. Also known as *Fusarium* head blight.

Scrapie A fatal disease of the nervous system in sheep and goats, characterized by pruritus, debility, and locomotor incoordination. It is caused by proteinaceous infectious particles called prions (USDA NAL).

Seafood Edible aquatic (freshwater or marine) organisms such as fish, shellfish, or seaweed that is used as food (USDA NAL).

Secondary metabolites Chemicals synthesized by organisms that do not appear to have a direct role in the organism's growth. Antibiotics and mycotoxins are examples of secondary metabolites.

Selectivity A term use in food testing to indicate how specific a test is for a select target. A high level of selectivity means a lower chance of cross-reaction with the wrong target.

Self-limited illness An illness that resolves on its own.

Sensitivity A term used in food testing to indicate the lower detectable limit of a testing procedure. The more sensitive a test is, the lower the concentration of target analyte or microbe may be detected.

Sepsis Systemic inflammatory response syndrome with a proven or suspected infectious etiology (USDA NAL).

Shiga-like toxin A toxin produced by certain pathogenic strains of *Escherichia coli* such as *Escherichia coli* O157. It is closely related to Shiga toxin produced by *Shigella dysenteriae* (USDA NAL).

Shiga toxins A class of toxins that inhibit protein synthesis by blocking the interaction of ribosomal RNA with peptide elongation factors. They include Shiga toxin which is produced by *Shigella dysenteriae* and a variety of shiga-like toxins that are produced by pathologic strains of *Escherichia coli* such as *Escherichia coli* O157 (USDA NAL). Same as verotoxins.

Single-cell protein Protein derived from a culture of single-celled organisms, used especially as a food supplement (Oxford Dictionaries).

Social accounting matrix (SAM) Represents flows of all economic transactions that take place within an economy (regional or national).

Spirulina Filamentous cyanobacteria that form tangled masses in warm alkaline lakes in Africa and Central and South America. Spirulina cyanobacteria dried and prepared as a food or food additive, being a rich source of many vitamins and minerals (Oxford Dictionaries).

Spore An environmentally resistant form of a cell that is dormant.

Spore germination When a spore comes out of dormancy and forms a biologically active cell.

Sporozoite A motile spore-like stage in the life cycle of some parasitic sporozoans (e.g. the malaria organism), which is typically the infective agent introduced into a host (Oxford Dictionaries).

Standard operating procedure (SOP) Written practices and procedures that are critical to producing safe food.

Staphylococcus aureus A Gram-positive, non-sporeforming, coccal bacteria that can produce a heat-stable enterotoxin.

Starvation Suffering or death caused by lack of food (Oxford Dictionaries).

Stationary phase The period of a growth curve after the log phase where multiplication and death are occurring at a similar rate and the population numbers remain steady.

Still birth Birth of an infant that died in the womb after 28 weeks of pregnancy.

Stochastic Having a random probability distribution or pattern that may be analysed statistically but may not be predicted precisely (Oxford Dictionaries).

Stochastic dominance Form of stochastic ordering where decisions can be based on best likely probabilities.

Storage fungi Fungi that grow in stored grain, spoiling quality and possibly safety.

Symbiosis Interaction between two different organisms living in close physical association, typically to the advantage of both (Oxford Dictionaries).

Symptom A physical or mental feature which is regarded as indicating a condition of disease, particularly such a feature that is apparent to the patient (Oxford Dictionaries).

Systemic infection An infection affecting the entire body rather than a single organ or body part.

T-2 toxin A potent mycotoxin produces by species of *Fusarium*.

Taenia A genus of flatworm that contains species that are parasites in humans.

Temperature abuse When foods are left in the temperature danger zone too long.

Temperature danger zone Temperatures between 40 and 140°F (4–60°C). This is the temperature range where most microbiological hazards for foods are able to grow and/or produce toxins.

Testing The use of laboratory-based methods to detect and/or measure substances or microorganisms in foods.

Thermoduric microorganisms Mesophiles that can survive periods of time at high temperatures.

Thermophile Microorganism that is hot loving and grows only at temperatures higher than 45°C. Thermophilic bacteria associated with foods have optimum temperature ranges for growth between 55 and 65°C.

Tightly coupled complex supply system In tightly coupled systems, performance standards are enforced and unambiguous (see Loosely coupled complex supply system).

Tightly coupled linear supply system In tightly coupled systems, performance standards are enforced and unambiguous (see Loosely coupled linear supply system).

Tissues Collections of differentiated cells, such as epithelium; connective tissue; muscles; and nerve tissue. Tissues are cooperatively arranged to form organs with specialized functions such as respiration; digestion; reproduction; movement; and others (USDA NAL).

Toxicity The intrinsic capacity of a chemical to produce injury.

Toxicoinfection A type of foodborne illness where large numbers of viable pathogenic bacteria are consumed and then release toxins that aid in colonization of the gastrointestinal tract of the host. Same as toxin-mediated infection.

Toxin A substance that has toxicity.

Toxin-mediated infection A type of foodborne illness where large numbers of viable pathogenic bacteria are consumed and then release toxins that aid in colonization of the gastrointestinal tract of the host. Same as toxicoinfection.

Toxoplasma gondii A protozoan parasite that can cause an infection known as toxoplasmosis.

Traceability The process of monitoring the movement of products from production to consumption and vice versa (USDA NAL). Trace back and trace forward are directions of traceability.

Transmissible spongiform encephalopathy A progressive condition that affects the brain and nervous systems of animals; prion disease.

Transmission The mechanism of transfer of pathogens to a host.

Traveler's diarrhea Diarrhea contracted as a result of eating contaminated foods while traveling. This is such a common affliction in travelers that it warrants its own terminology.

Trichotecenes Usually 12,13-epoxytrichothecenes, produced by Fusaria, *Stachybotrys*, *Trichoderma* and other fungi, and some higher plants. They may contaminate food or feed grains, induce emesis and hemorrhage in lungs and brain, and damage bone marrow due to protein and DNA synthesis inhibition (USDA NAL).

Type I error A false positive.

Type II error A false negative.

Typhoid An infectious bacterial fever with an eruption of red spots on the chest and abdomen and severe intestinal irritation (Oxford Dictionaries).

Typhoid Mary A transmitter of undesirable opinions or attitudes. The nickname of Mary Mallon (died 1938), an Irish-born cook who transmitted typhoid fever in the US (Oxford Dictionaries).

United States Department of Agriculture (USDA) A department of the United States government that manages various programs related to food, agriculture, natural resources, rural development, and nutrition.

Usability Used to describe the user friendliness of analytical methods.

Vacuum packaging The packaging of processed products in which air is removed and a vacuum is formed. Carbon dioxide gas or nitrogen gas may be introduced into the package to help eliminate the oxygen. Removal of the air retards growth of aerobic bacteria (USDA NAL).

Variant Creutzfeldt–Jakob disease (vCJD) A human disease thought to be caused by the same prion that causes bovine spongiform encephalopathy.

Vectors Invertebrates or non-human vertebrates which transmit infective organisms from one host to another (USDA NAL).

Vegemite A type of savoury spread made from concentrated yeast extract (Oxford Dictionaries).

Vegetable Any part of a plant that is commonly eaten by humans as food, but is not considered to be a culinary fruit, nut, herb, spice or grain (USDA NAL).

Vegetative cell A cell that is actively growing rather than forming spores.

Verotoxins A class of toxins that inhibit protein synthesis by blocking the interaction of ribosomal RNA with peptide elongation factors. They include Shiga toxin which is produced by *Shigella dysenteriae* and a variety of shiga-like toxins that are produced by pathologic strains of *Escherichia coli* such as *Escherichia coli* O157 (USDA NAL). Same as shiga toxins.

Viable Capable of surviving or living successfully, especially under particular environmental conditions (Oxford Dictionaries).

Viable but nonculturable Bacteria that are in a state of very low metabolic activity and do not divide, but are alive and have the ability to become biologically active.

Vibrio Gram-negative, small, curved, rod-shaped bacteria that may cause infections known as vibriosis. *Vibrio cholerae* causes a toxin-mediated infection known as cholera.

Virotypes Populations of the same species which differ by their pathogenic capability (USDA NAL). Same as pathotypes or pathovars.

Virulence The degree of pathogenicity within a group or species of microorganisms or viruses as indicated by case fatality rates and/or the ability of the organism to invade the tissues of the host. The pathogenic capacity of an organism is determined by its virulence factors (USDA NAL).

Virus An infective agent that typically consists of a nucleic acid molecule in a protein coat, is too small to be seen by light microscopy, and is able to multiply only within the living cells of a host (Oxford Dictionaries).

Vomit Eject matter from the stomach through the mouth (Oxford Dictionaries).

Vomitoxin The most abundantly produced mycotoxin within the family of mycotoxins known as trichothecenes. Also known as deoxynivalenol (DON).

Water activity $a_w = p/p_o$, where p is the vapor pressure of the solution (food) and p_o is the vapor pressure of the solvent (water). The amount of water available for microbial use. The a_w of pure water = 1.0.

Whistleblowing The reporting of observed or suspected professional misconduct or incompetence to appropriate authorities or to the public (USDA NAL).

World Trade Organization (WTO) Global international organization dealing with the rules of trade between nations.

Xerophile (Of a plant or animal) adapted to a very dry climate or habitat, or to conditions where moisture is scarce (Oxford Dictionaries). Dry loving.

Yersinia Gram-negative, pshychrotrophic, straight or curved rod-shaped bacteria that may cause an infection known as yersiniosis.

Zoonotic diseases or zoonoses Diseases of non-human animals that may be transmitted to humans or may be transmitted from humans to non-human animals (USDA NAL).

Index

Page numbers in **bold** refer to glossary entries, those in *italics* refer to figures.